WHAT EVERY ENGINEER SHOULD KNOW ABOUT

ETHICS

WHAT EVERY ENGINEER SHOULD KNOW

A Series

Editor

William H. Middendorf

Department of Electrical and Computer Engineering
University of Cincinnati
Cincinnati, Ohio

1. What Every Engineer Should Know About Patents, *William G. Konold, Bruce Tittel, Donald F. Frei, and David S. Stallard*
2. What Every Engineer Should Know About Product Liability, *James F. Thorpe and William H. Middendorf*
3. What Every Engineer Should Know About Microcomputers: Hardware/Software Design, A Step-by-Step Example, *William S. Bennett and Carl F. Evert, Jr.*
4. What Every Engineer Should Know About Economic Decision Analysis, *Dean S. Shupe*
5. What Every Engineer Should Know About Human Resources Management, *Desmond D. Martin and Richard L. Shell*
6. What Every Engineer Should Know About Manufacturing Cost Estimating, *Eric M. Malstrom*
7. What Every Engineer Should Know About Inventing, *William H. Middendorf*
8. What Every Engineer Should Know About Technology Transfer and Innovation, *Louis N. Mogavero and Robert S. Shane*
9. What Every Engineer Should Know About Project Management, *Arnold M. Ruskin and W. Eugene Estes*
10. What Every Engineer Should Know About Computer-Aided Design and Computer-Aided Manufacturing: The CAD/CAM Revolution, *John K. Krouse*
11. What Every Engineer Should Know About Robots, *Maurice I. Zeldman*
12. What Every Engineer Should Know About Microcomputer Systems Design and Debugging, *Bill Wray and Bill Crawford*
13. What Every Engineer Should Know About Engineering Information Resources, *Margaret T. Schenk and James K. Webster*
14. What Every Engineer Should Know About Microcomputer Program Design, *Keith R. Wehmeyer*
15. What Every Engineer Should Know About Computer Modeling and Simulation, *Don M. Ingels*

16. What Every Engineer Should Know About Engineering Workstations, *Justin E. Harlow III*
17. What Every Engineer Should Know About Practical CAD/CAM Appli cations, *John Stark*
18. What Every Engineer Should Know About Threaded Fasteners: Materials and Design, *Alexander Blake*
19. What Every Engineer Should Know About Data Communications, *Carl Stephen Clifton*
20. What Every Engineer Should Know About Material and Component Failure, Failure Analysis, and Litigation, *Lawrence E. Murr*
21. What Every Engineer Should Know About Corrosion, *Philip Schweitzer*
22. What Every Engineer Should Know About Lasers, *D. C. Winburn*
23. What Every Engineer Should Know About Finite Element Analysis, *edited by John R. Brauer*
24. What Every Engineer Should Know About Patents: Second Edition, *William G. Konold, Bruce Tittel, Donald F. Frei, and David S. Stallard*
25. What Every Engineer Should Know About Electronic Communications Systems, *L. R. McKay*
26. What Every Engineer Should Know About Quality Control, *Thomas Pyzdek*
27. What Every Engineer Should Know About Microcomputers: Hardware/Software Design, A Step-by-Step Example. Second Edition, Revised and Expanded, *William S. Bennett, Carl F. Evert, and Leslie C. Lander*
28. What Every Engineer Should Know About Ceramics, *Solomon Musikant*
29. What Every Engineer Should Know About Developing Plastics Products, *Bruce C. Wendle*
30. What Every Engineer Should Know About Reliability and Risk Analysis, *M. Modarres*
31. What Every Engineer Should Know About Finite Element Analysis: Second Edition, Revised and Expanded, *edited by John R. Brauer*
32. What Every Engineer Should Know About Accounting and Finance, *Jae K. Shim and Norman Henteleff*
33. What Every Engineer Should Know About Project Management: Second Edition, Revised and Expanded, *Arnold M. Ruskin and W. Eugene Estes*
34. What Every Engineer Should Know About Concurrent Engineering, *Thomas A. Salomone*
35. What Every Engineer Should Know About Ethics, *Kenneth K. Humphreys*

ADDITIONAL VOLUMES IN PREPARATION

WHAT EVERY ENGINEER SHOULD KNOW ABOUT ETHICS

Kenneth K. Humphreys, P.E., C.C.E.
Consulting Engineer
Granite Falls, North Carolina

MARCEL DEKKER, INC. NEW YORK • BASEL

ISBN: 0-8247-8208-9

This book is printed on acid-free paper.

Headquarters
Marcel Dekker, Inc.
270 Madison Avenue, New York, NY 10016
tel: 212-696-9000; fax: 212-685-4540

Eastern Hemisphere Distribution
Marcel Dekker AG
Hutgasse 4, Postfach 812, CH-4001 Basel, Switzerland
tel: 41-61-261-8482; fax: 41-61-261-8896

World Wide Web
http://www.dekker.com

The publisher offers discounts on this book when ordered in bulk quantities. For more information, write to Special Sales/Professional Marketing at the headquarters address above.

Current printing (last digit):
10 9 8 7 6 5 4 3 2 1

PRINTED IN THE UNITED STATES OF AMERICA

Preface

This book is intended to be a practical guide and textbook for students and practicing engineers and to explain in succinct terms the codes of ethics of the engineering profession, legal requirements of ethical practice, and ethical dilemmas faced by engineers. These dilemmas are illustrated by case studies drawn from the decisions of the National Society of Professional Engineers' Board of Ethical Review and other sources. A few of the more spectacular news-making events of recent years are discussed (e.g., the Challenger Disaster and the Hyatt-Regency Hotel collapse in Kansas City), but, unlike some other books on the subject, these are not the major focus of the book. These headline-making issues, while of great public interest, rarely involve the types of ethical dilemmas commonly encountered by engineers. *What Every Engineer Should Know About Ethics* concentrates instead on the types of problems engineers more commonly encounter on a daily basis. The book is not a philosophical treatise, rather it is intended to be a practical reference work, both for the practicing engineer and the engineering student. It is designed for ready use as a text for undergraduate and graduate engineering ethics courses and as a supplemental text for courses which incorporate units on engineering ethics.

I extend my sincere appreciation to the various individuals and engineering societies that have contributed to the preparation of this volume. In particular, I thank Mr. Russell C. Lindsay, Dr. Leonard Ferenz, and Dr. Peter J. Rutland for their significant contributions. Mr. Lindsay is a registered professional engineer and is President of PARC Engineering Associates, Asheville, North Carolina. Dr. Ferenz is a Faculty Associate with the Center for Professional Ethics, University of North Carolina at Charlotte, Charlotte, North Carolina. Dr. Rutland hails from the opposite side of the word as Chief Executive Officer, Waiariki Institute of Technology, Rotarua, New Zealand.

Within the book, I have acknowledged the professional societies that have contributed to the book in various ways. Two societies, however, deserve particular mention---the National Society of Professional Engineers (NSPE) and the Professional Engineers of North Carolina (PENC). NSPE has granted permission for me to freely draw upon and include case studies from their Board of Ethical Review and other NSPE documents. I am deeply grateful to NSPE for their support.

I am chairman of the PENC Ethics Steering Committee and am a regular contributor of articles to the "Ethics Corner" section of its magazine, *the Professional Engineer, The Magazine of North Carolina Engineering.* Mr. Lindsay is a member of the Ethics Steering Committee of PENC and also contributes articles to the magazine. PENC has given its full support to the development of this book and has permitted the use of much material from these articles in the book. The material is woven throughout the text and accordingly is not always specifically credited to PENC. However, it is reasonable to say that were it not for PENC, the idea for this book would never have come about and, without its agreement to let me use the material from the articles, the job of completing the book would have been much more difficult.

Last, I want to acknowledge the invaluable support and critical editorial eye of my wife Betsy. Without her encouragement I would never have begun this effort, and without her support I would not have been able to complete it.

Kenneth K. Humphreys, P.E., C.C.E.

Contents

1

Problem-Solving in Engineering Ethics*

CASE STUDY

An engineer with a firm specializing in electronics manufacturing is asked by management to design and develop a new generation of radar detectors for private consumption. The newest generation of "stealth" detectors is to alert drivers to the presence of police radar and laser traps while at the same time "cloaking" use of the detectors.

Now, let us suppose our engineer has misgivings about developing such a device. His reasoning is as follows: Radar detectors, as the name implies, are most often used by drivers to detect "radar." Radar, of course, is any of a group of electronic devices used by the highway patrol to gauge the speed of traffic on the roadways. Drivers generally use such detectors to alert them to the use of radar. Because radar detectors are used by drivers primarily to exceed posted speed limits, which is an illegal act, some states have made their use illegal. Thus, in order to discover the use of "illegal" radar detectors, law enforcement officers employ electronic devices designed to detect the use of radar detectors. The device that our engineer is being asked to develop is one that will alert the user that such countermeasures are being used without it being detected.

Now, imagine further that our engineer believes that speed limits are a reasonable limitation on personal freedom because they help secure the safety of other persons traveling on or nearby the roadways, such as pedestrians or the drivers and occupants of other vehicles. Let us also suppose our engineer believes that the device he is being asked to develop will be used almost exclusively to break what he considers to be just laws. Troubled, he expresses his concerns to management.

---

* By Dr. Leonard Ferenz, Center for Professional and Applied Ethics, The University of North Carolina at Charlotte, Charlotte, North Carolina

Management tells our engineer that there are significant profits in such a device and that he should get to work on the project or find another job. The dilemma for our engineer, then, is whether to serve his own and his employers' interests by doing what he is told and, thereby, keep his job or serve the public interest by refusing to participate in the development of a device that will help enable activities that are a threat to public safety.

INTRODUCTION

Similar dilemmas could face engineers being asked to cover-up, ignore, or under-report the existence of toxic wastes, structural defects, design flaws, or any other potential health or safety hazards. Under such circumstances, what is the "right" thing to do?

The solution in the above case may seem obvious. But the reader may be assured that whatever his or her opinion at the present time, there will be others with a different opinion about what to do in this case. Indeed various types of professionals, including engineers, often express significant differences of opinion when faced with cases requiring an ethical solution. The objective before us in this chapter, therefore, is to provide a framework for reconciling differences of opinion as we address the question, "What is the right thing to do?" in circumstances involving ethical issues in the engineering profession.

There are many ways to consider the question, "What is the right thing to do?" From an economic perspective, the right thing to do is whatever is the most profitable. From a personal interest perspective, the right thing to do is whatever maximizes one's own well being. There may be religious perspectives, political perspectives, social perspectives, etc. However, the perspective to be considered in this chapter is the moral, or ethical, perspective. We want to know what is the morally correct, or right, thing to do when faced with situations involving questions of right and wrong, good and evil, virtue and vice.

What we will not do in this chapter, however, is to lay out a particular set of moral rules or promote a particular moral viewpoint. This chapter does not attempt to provide the "right" answers to specific moral problems. What we are concerned with here is a practical and workable method for dealing with moral problems in specific cases. In medicine, moral decision-making involving specific persons in real-life circumstances is called "clinical ethics" since it generally involves ethical decisions involving patients in a clinical setting. In this chapter, we will be developing a method for solving ethical problems in specific cases that is the engineering equivalent of clinical ethics in medicine. We will refer to this simply as "case ethics."

Our chapter begins with a short list of ethical terms and definitions. This list is designed to equip the reader with a decent understanding of ethical terms, not only for the benefit of this discussion but also to facilitate meaningful moral deliberation and discussion with others. In this list I have tried to avoid much of the jargon that occurs in academia and have concentrated on terms that the reader would most likely run across in the literature on professional ethics.

Next, we look at various options available to engineers to address ethical issues in their professional practices. Foremost among these is the *Code of Ethics for Engineers*, developed by the National Society of Professional Engineers. There are a number of concerns that have been raised with regard to such *professional codes of ethics* in general and this one in particular. We will take a brief look at some of the more important of these concerns.

Then, we will look at a problem-solving method for case analysis in ethics that has had tremendous success in various professional settings. It is a method that can be used by individuals alone or in consultation with others. Finally, we will look at the role of ethics committees in case consultation, policy formation, and review, and as a forum for ethical deliberation. The use of ethics committees as an approach to solving ethical problems arising in the professions has several important implications. We will take a brief look at these and then offer some concluding thoughts.

ETHICAL TERMS AND DEFINITIONS

In contemporary literature, the term "ethics" often refers to reflective and theoretical perspectives of right and wrong, what dictionaries call "moral philosophy." It also refers to the system or code of morals practiced by a particular person, group, or profession. The term "morality" refers generally to actual principles of conduct practiced by individuals or groups of individuals or to "ethics." Because of the obvious overlap in the meaning of these terms, we generally find both being used interchangeably in the literature. For this reason, the words "morality" and "ethics" will be treated as synonyms in this chapter.

In their broadest and most familiar meaning, morality and ethics are concerned with many forms of belief about good and bad, right and wrong, appropriate and inappropriate human behavior, rights, virtue, and vice. Morality and ethics are studies of what we *ought* to do and how we *ought* to behave from a moral viewpoint, as opposed to an economic, religious, political, or prudential viewpoint. From these perspectives, what we ought to do may be very different from what we ought to do from a moral perspective. For example, if I want to get to an important interview on time, it might be prudent for me to exceed the speed limits as I drive to my appointment. I ought, in this case, to speed. However, if it's morally wrong to speed, then the moral thing to do is not speed. In this case, as is often the case, there is a difference between what I ought to do morally, the so-called *moral ought*, and what I ought to do to further some non-moral goal (such as an economic goal), which is sometimes referred to as the *prudential ought*.

The kinds of situations that are particularly challenging in ethics are those that involve an *ethical dilemma*. An ethical dilemma exists whenever moral reasons or considerations can be offered to support two or more opposing courses of action. For example, respect for individual self-determination could be offered as a moral reason to support a person's decision not to wear seatbelts while respect for the value of human life might be used to support, or justify, mandatory seatbelt laws.

An ethical dilemma is not the same as an ethical issue. An *ethical issue* is a general topic or problem involving moral rules, principles, and norms. For example, physician-assisted suicide is an ethical issue in medicine; vendor/supplier policies, perks, and gifts are ethical issues in business; and weapons research is an ethical issue in engineering. The reason these topics are ethical issues, of course, is because of important differences of opinion and disagreement concerning what is morally correct.

Moral Rules, Principles, and Duties

Much of the discussion in ethics involves the use of some key concepts and principles. To illustrate, consider whether an individual should have the right to refuse life-saving medical treatments. Today, one could argue that patients do have a legal right to accept or refuse life-saving medical treatment, as recognized in such legislation as the Patient Self-Determination Act of 1990 (1). But many would argue that what makes this legislation defensible from a moral point of view is the overriding importance of a moral principle, the so-called *principle of autonomy*. According to this principle, the actions and decisions of *autonomous* persons ought to be respected. Thus, the reason patients have a right to accept or refuse medical treatment is because the principle of autonomy demands that we respect the actions and decisions of autonomous persons.

On the other hand, many would argue that we have moral obligations to promote the good of others and to preserve human life whenever possible. The *principle of beneficence* demands that we promote the good of others. The good, as is often defined when discussing this principle, includes the preservation of life.

Now, consider an ethical dilemma involving these two principles. Imagine that a practicing Jehovah's Witness is brought into the emergency room after sustaining injuries in an automobile accident. He is an adult, he is bleeding to death, and he refuses to accept a life-saving blood transfusion. Assuming the patient is clear-headed and capable of making an informed decision, what should the emergency room physician do? Should the physician respect the patient's refusal, as demanded by the principle of autonomy, or transfuse the patient against his will, as demanded by the principle of beneficence? As you can see, this case poses an ethical dilemma because moral reasons or considerations can be offered to support two opposing courses of action. We can describe that dilemma as a conflict between the moral principles of autonomy and beneficence.

Notice, however, that being able to describe an ethical dilemma in terms of such principles as autonomy and beneficence does not solve the ethical dilemma. It is not obvious in this case, for example, whether respect for the autonomy of the Jehovah's Witness should override our obligations of beneficence or vice versa. In fact, we often find that autonomy will override beneficence in one case but not in another depending upon the circumstances of the case. Because there is no apparent hierarchy or ranking of moral principles and their corresponding duties, they are generally considered to be *prima facie*; that is, "at first sight" each principle is binding at all times in all cases. It is the particular circumstances of the case that will cause one of these principles to override another.

While describing an ethical dilemma in terms of its competing moral principles will not solve the dilemma, it does help identify important moral values and interests that are at stake. To this end, the following list includes a brief description of the more commonly used moral principles and duties in contemporary discussions of ethics. Note that each of these principles, or action-guides, generates corresponding duties, listed immediately to the right of each principle. Also be aware that the order in which these principles occur in the following list should not be considered as an indication of moral rank or priority since each principle is *prima facie*.

Principle	*Corresponding Duty*
the principle of autonomy	respect the autonomy of others
the principle of nonmaleficence	do not inflict harm on others
the principle of beneficence	promote the good of others
the principle of justice	give others what is due or owed to them; give others what they deserve
the principle of truth-telling	disclose all relevant information honestly and intelligibly; do not intentionally deceive
the principle of promise-keeping	be faithful to just agreements; honor contracts

The *principle of autonomy* requires that we respect the autonomy of others. It asserts a right of noninterference and correlatively an obligation not to constrain or interfere with the autonomous actions of others. It is based on the idea that persons possess "autonomy"; that is, that they are free, self-governing, and self-determining beings. An individual is *autonomous* in the absence of internal or external constraints that would compromise the ability to act voluntarily toward a chosen course of action or to fulfill a chosen life-plan.

The *principle of nonmaleficence* requires that persons do not intentionally, or deliberately, do harm to others. According to the nineteenth century philosopher John Stuart Mill, we can "cause evil" to others by both our *actions* and our *inactions* (2). General usage of this principle refers only to harms caused through actions, that is, positive acts, or *commissions*, and not through our inactions, or *omissions*. For example, the duty of nonmaleficence requires that we forebear from inflicting harm or injury on others by stealing from them or assaulting them, but it does not require that we rescue a person who is drowning. Thus, the principle of nonmaleficence is sometimes referred to as a *negative injunction* since it tells us what not to do.

The *principle of beneficence* is based upon the concept of *beneficence*, which involves producing benefits and promoting the good for others. The principle of beneficence is a *positive injunction* that imposes obligations both to provide benefits and to prevent and remove harms. According to a primary text in medical ethics, the principle of beneficence "asserts the duty to help others further their important and legitimate interests" and, more specifically, it asserts "the duty to confer benefits and actively to prevent and remove harms" (3). Philosopher William Frankena combines obligations of nonmaleficence with those of beneficence. According to Frankena (4), the principle of beneficence imposes four duties:

1. One ought not to inflict evil or harm (what is bad)
2. One ought to prevent evil or harm
3. One ought to remove evil
4. One ought to do or promote good

According to either interpretation, the principle of beneficence requires the prevention of evil through our inactions. Thus, we would be required according to this principle to rescue someone who is drowning.

The *principle of justice* generally requires doing what is *fair*, or *just*. For example, according to the philosopher John Rawls, "justice as fairness" requires that we promote the good of the least well-off in society prior to the good of those who are better off until all enjoy equal shares of the good. According to Rawls, inequities in the distribution of goods are justifiable only to the extent that they further the interests of the least well-off in society (5). So, for example, if children suffering from life-threatening genetic diseases were considered the least well-off in society, they would have the prior claim on society's resources. Thus we would be justified in rewarding geneticists and biochemists disproportionately to other professionals in society if their work was the most likely to improve the well being of these least well-off members of society.

Rawls' is just one version of a conception of justice called *egalitarianism*. Other competing conceptions of justice include various forms of *utilitarianism* (which focuses on securing "the greatest good for the greatest number" of persons in society), *libertarianism* (a kind of *laissez faire* approach to economic distributions), *socialism*, *Marxism*, and various *equality of opportunity* models. While the social and economic implications of each of these models differ significantly, they all involve what persons *deserve* or can legitimately claim on the basis of certain morally relevant properties, such as age, merit, productivity, social status, economic status, etc.

The *principle of honesty*, or *truth-telling*, obviously requires telling the truth when asked but it can also be interpreted more stringently. In which case, truth-telling also requires full and impartial disclosure of any information that could be considered legitimately important to others.

The *principle of fidelity*, or *promise-keeping*, recognizes an obligation to honor *just* agreements and to keep promises freely entered into and deliberately made.

Returning to the case at the beginning of this chapter, we can now use these principles to describe the ethical dilemma that faces our engineer. On the one hand, our engineer will have obligations of fidelity to honor the terms of his employment contract, which may require that he accept assignments suited to his expertise. According to the principle of beneficence, he also has obligations to promote the good of others, including his employers, the people he works with, and certainly his own family. If the good of these others is at risk if he should quit or if the financial security of his family is at risk if he loses his job, then beneficence would enjoin him or her to accept the assignment.

The principle of beneficence, however, asks us to consider the good of *all* others, not just family or employers. Considering the good of the public in general, our engineer may feel that the good of many "others" would be best served by refusing to work on a device whose sole purpose, he believes, would be to allow law-breakers to avoid detection when speeding. Moreover, the principle of autonomy might be invoked here to defend the position that our engineer's decision should be respected. However, the principle of autonomy also tells us that we should respect the freedom of others to be self-determining. One could argue along these lines that people should be free to purchase whatever products they like as long as the purchasing in itself presents no risk of harm to others. One might argue that how the product is used, or *misused*, later is the business of the purchaser, not the seller and certainly not the designer.

Familiarity with this list of principles and duties can provide a very effective way to describe various ethical considerations. However, that alone will not solve the ethical dilemma. In fact, we have even seen that the very same principles can be offered to support opposing courses of action. In view of the fact that there are no obvious solutions to such ethical difficulties, we should consider the professional codes of ethics that govern engineering practices.

PROFESSIONAL CODES OF ETHICS FOR ENGINEERS

The *Model Guide for Professional Conduct* developed by the American Association of Engineering Societies (AAES) and the *Code of Ethics for Engineers* of the National Society of Professional Engineers (NSPE) are the two dominant codes affecting the engineering profession. Both appear in the Appendix along with several other codes developed by other engineering-related organizations. The AAES's *Model Guide* includes 10 briefly stated "Canons of Professional Conduct." The *Code* developed by the NSPE is certainly longer and more detailed. It includes six "Fundamental Canons," five lengthy "Rules of Practice," and nine equally lengthy "Professional Obligations." The NSPE, with its 54 constituent state and territorial engineering societies, is the umbrella organization that is concerned with non-technical and professional matters for engineers. The principles of the NSPE *Code* are generally mandated by law or regulation for those persons registered to practice under current state registration laws. Moreover, in the 1950s, the NSPE established a Board of Ethical Review whose job it is to review cases submitted to the Board by engineers, public officials, ethicists, educators, and members of the general public.

In 1988, the NSPE also created the National Institute for Engineering Ethics (NIEE). NIEE is now an independent not-for-profit educational corporation. NIEE's mission is to provide opportunities for ethics education and to promote the understanding and application of ethical processes within the engineering profession and with the public.

Considering the general inclusion of the NSPE's more detailed *Code of Ethics* in state and territorial laws and/or regulations governing the practice of engineering, our engineer in the opening case of this chapter would be well advised to consult the NSPE's *Code* for moral guidance. Indeed, there he would find that the first fundamental canon that guides engineers in "the fulfillment of their professional duties" is to "Hold paramount the safety, health and welfare of the public" (6).

The word "paramount" is generally understood to mean "the highest in rank," pre-eminent, or supreme. This suggests, then, that engineers are expected to act in such a way that their professional conduct will promote, preserve, and protect public safety, health, and welfare and that this objective should be the overriding concern of all engineering professionals. This would seem to suggest, then, that the engineer in our opening case *must,* if he is to comply with the requirements of the *Code*, refuse to work on the project, even at the risk of losing his job.

This interpretation appears to be consistent with the opinions that have been offered by the NSPE's Board of Ethical Review in their consideration of several cases. The dominant theme that has emerged in the interpretation of this first canon is that engineers are not only allowed, but are required, to refuse to participate in projects that they believe jeopardize public health and safety. For example, in a decision involving a group of engineers who believed that certain machinery was unsafe, the Board "determined that the engineers were ethically justified in refusing to participate in the processing or production in question. We recognized in that case that such action by the engineers would likely lead to the loss of employment" (7). This statement recognizes that the engineers are justified in refusing to participate in a questionable project but does not state that they would be *required* to refuse. The Board does take this additional step, however, in an opinion involving an engineer faced with "blowing the whistle" where they concluded that "the *Code* only *requires* [emphasis added] that the engineer withdraw from a project and report to proper authorities when the circumstances involve endangerment to the public safety, health and welfare" (8). In another case involving an engineer employed as the City Engineer/Director of Public Works and her concerns about the management of waste water from ponds containing domestic waste, the Board argues that "the engineer has not merely an 'ethical right' but has an 'ethical obligation' to report the matter to the proper authorities and withdraw from further service on the project." The Board acknowledges "the basic reality" that "the engineer who makes the decision to 'blow the whistle' will in many instances be faced with the loss of employment" (9). As suggested by these comments, it appears that the NSPE's Board takes very seriously the idea that public safety is paramount.

What this suggests in the present case is that our engineer is ethically *required* to refuse to participate in this project even at the risk of losing his job.

MERITS AND DRAWBACKS OF PROFESSIONAL CODES OF ETHICS

Merits

One considerable advantage of professional codes of ethics is that they define what are considered within the profession to be the moral responsibilities of practitioners in the field. They do this by offering principles and rules of ethical behavior tailored to a particular profession. These guidelines help establish what can be termed the *role-specific duties* of members of the profession.

An additional benefit provided by the NSPE Board of Ethical Review and other professional organizations is the periodic publication of advisory opinions on non-technical matters that pose ethical problems. Advisory opinions are published for instructive purposes and to aid in the interpretation and application of the *Code's* ethical guidelines in these and other real-life situations. In addition to requests by mail or phone, anyone with Internet access can review the advisory opinions of the Board.

Professional codes of ethics may also promote some uniformity in professional conduct and the manner in which ethical issues are addressed. For example, the NSPE's The Board of Ethical Review is composed of engineers from various disciplines who share a similar interest in encouraging "adherence to the highest principles of ethical conduct" on behalf of the public, clients, employers, and the profession (10). If members of the profession do adhere to these principles or if their conduct is generally reflected in the judgments and opinions expressed by the Board, then we would expect to see some uniformity in the professional conduct of engineers generally. Not only might this lead to some uniformity in the way professionals address ethical matters, but it might also be seen as helping to establish a bond among professionals in the field.

Drawbacks and Criticisms

Lack of Specificity and Problem-Solving Capability

A common criticism of professional codes of ethics is that they tend to be too general, and therefore provide little guidance in specific circumstances and situations involving true moral dilemmas. For example, AAES's *Model Guide for Professional Conduct* requires that "Engineers are responsible for enhancing their professional competence throughout their careers and for encouraging similar actions by their colleagues" (11). But just what would be involved in doing this? What is actually expected of an engineer? Is the AAES suggesting that engineers should take continuing education classes, subscribe to journals, participate in seminars and conferences, etc.? And how often?

The NSPE's *Code of Ethics for Engineers* also seems rather general. For example, it requires that "Engineers shall seek opportunities to participate in civic affairs; career guidance for youths; and work for the advancement of the safety, health and well-being of their community." But the *Code* does not tell us what *kinds* of opportunities and how often, or how much of one's time is to be devoted to this pursuit. In addition to being vague, this may be unrealistic and overly demanding depending on how much engineers are expected to do to fulfill this requirement.

The NSPE's *Code* also requires, for example, that "Engineers shall avoid the use of statements containing a material misrepresentation of fact or omitting a material fact" (12). This injunction may be applied to advertising and would clearly prohibit false and misleading advertisements. But what it does not tell us is just what kinds of statements or advertisements are deceptive or misleading. We are all quite familiar with how successfully we can deceive others, not only by lying, but also by choosing not to disclose information. According to this rule, engineers should not make false statements, but it is not clear what engineers are required to disclose nor what would be considered misleading.

The more important criticism regarding the generality of professional codes of ethics, however, is that they do not provide much guidance resolving important moral conflicts. For example, the NSPE *Code* requires that "Engineers shall not disclose, without consent, confidential information concerning the business affairs or technical processes of any present or former client or employer, or public body on which they serve" (13). Now, consider an engineer who is hired by a client to do an assessment of toxic waste disposal at the client's industrial facility. The engineer discovers that leaks from waste storage drums violate certain federal or state laws. The engineer reports this to the client who then thanks her, pays the consulting fee, and then ignores the report. Should the engineer report these violations to the proper authorities or remain silent? The engineer should clearly not remain silent according to the *Code's* statement that "Engineers having knowledge of any alleged violation of the *Code* shall report thereon to appropriate professional bodies, and when relevant, also to public authorities, and cooperate with the proper authorities in furnishing such information or assistance as may be required." Obviously, if the "alleged violation" was committed by a client, then the engineer is required to "blow the whistle" even though practices like the disposal of toxic waste are clearly among the "business affairs" of the client. The difficulty in interpreting these two requirements in the *Code* is that we really do not know what is meant by "confidential information." If waste management results are not confidential business, then there is no conflict between these requirements. But it is not clear in the *Code* itself what kinds of information are confidential.

In this example, there appears to be an internal conflict in the requirements of the *Code* with no clear indication of which rule is overriding. Unfortunately, there is no clear guidance on how to resolve the moral conflict between fidelity to one's client and obligations to the state. These concerns about a lack of specificity and problem-solving ability are among the primary reasons many critics argue that professional codes of ethics are inadequate to guide professional conduct.

Narrowness of Scope and Lack of Moral Commitment

Another criticism of professional codes of ethics is that they are too narrow in scope and lack moral commitment. The AAES's *Model Guide* for example, insists that "Engineers act in accordance with all applicable laws..." (11). At first sight, this seems not only prudent but reasonable. Yet, some would argue that when moral obligations exceed or conflict with one's legal obligations, one is morally bound to act in a militant and possibly even illegal manner. Consider these examples from medicine:

1. For some patients, marijuana is the most effective pain-control substance with the fewest adverse side effects. In light of current legal prohibitions against the use and sale of marijuana, should physicians recommend marijuana to their patients or possibly even help them procure it?
2. Some dying patients and their physicians agree that death would be in the patient's best interest as a release from uncontrollable pain, discomfort, and decline. Though not illegal, physician-assisted suicide presently is not legal in the U.S. Should physicians assist a patient in committing suicide when their condition is incurable and they are in uncontrollable pain?
3. Consider abortions. Even though abortions are presently legal, perhaps physicians should refuse collectively to perform abortions. On the other hand, perhaps if abortion were illegal, physicians should agree collectively to perform illegal abortions.

The point here, of course, is that questions concerning what professionals, such as physicians or engineers, "ought" to do *morally* as professionals is ultimately a moral matter, not a legal one. Thus, although there are people who would argue that morality does indeed make obedience to the law, whatever that might happen to be, the professional's preeminent *moral* obligation, most critics will argue that when morality and the law do conflict, one is morally bound to follow the dictates of morality.

Consequently, if there are occasions when morality does demand professional conduct that conflicts with or surpasses the demands of the law, then professional codes of ethics that require compliance with standards that fall within the parameters of our present legal statutes may be criticized as lacking a moral commitment.

Lack of Legal Authority

A related criticism of professional codes of ethics is that they lack legal authority. Organizations, such as the AAES or NSPE, are not federal or state law enforcement agencies. Often, the extent of the punitive measures that can be taken are limited to reprimands, publication of the transgressors name in a professional circulation, or suspension or termination of membership in the organization. The rules of the NSPE's *Code*, for example, are generally enforceable only among NSPE members, through affiliated state societies whose primary function is the administration of disciplinary action. It should be noted, however, that in some instances portions of the NSPE's *Code* have been incorporated into state laws and, in those instances, those portions are legally enforceable. This is discussed further in Chapter 5.

Lack of Objectivity

Another criticism of professional codes of ethics is that they may be biased toward members of the profession. For example, the judgments of an ethics board or committee made up exclusively or predominantly by members of the profession may not reflect the moral sentiments and expectations of persons generally. If they do not, then these judgments are likely to carry very little weight outside the profession. There may be serious concerns about the impartiality and objectivity of a committee or board drawn predominantly from members of the profession. By analogy, we would have similar doubts about the impartiality and objectivity of a jury in a murder case made up exclusively of members of the defendant's family.

Lack of Moral Authority

The question of moral authority points toward a fundamental concern about the justification basis for professional codes of ethics in general. Are any of these codes grounded in universally, or even generally, accepted moral standards? However, even though there are discrepancies among codes of ethics, this does not mean that none of them are grounded in such standards. The question we need to ask is whether it just so happens that some code available to engineers, such as the NSPE's, is one that is sufficiently grounded. If it is, then we would have a compelling reason to accept that code. If it is not, we would be compelled to look elsewhere for moral guidance.

ALTERNATIVE APPROACHES TO MORALITY

Models of Moral Responsibility

In addition to the *professional model* of moral responsibility expressed in various professional codes of ethics, there are at least five other models of moral responsibility worth considering: the subjective, or egoistic, model; the legal model; the societal model; the religious model; the so-called Natural Law model; and, finally, a philosophical model.

The Subjective, or Egoistic, Approach to Ethics

Many professionals and nonprofessionals think it is enough to act in a manner that they personally believe to be ethical for their conduct to be, in fact, morally correct. According to this approach, the moral responsibilities of engineers are those recognized individually by each member of the profession. Awareness, sensitivity, and appeals to conscience, however, will not eliminate or resolve conflicts among persons equally concerned with acting ethically. The debate over physician-assisted suicide among health professionals is a striking example of disagreement among members of the same profession—if not equally, at least enthusiastically—committed to doing what they believe is the right thing. Because of the extreme and often conflicting diversity of opinion in matters of morality, allowing each individual member of a profession to define for themselves their own code of ethics would almost certainly lead to as many different codes of ethics as there are members of the profession. It is precisely because of disagreement at the level of personal opinion or personal intuition that the subjective approach to questions of morality is generally unacceptable.

There is an additional, deeper, concern about the kind of reasoning that instructs us to be ethical for the sake of prudence and our own self-interest. That objection is simply that there would be no reason to behave ethically if we could better serve our own interests by acting unethically.

The Legal Approach to Ethics

It is sometimes argued that our present laws and statutes establish the appropriate parameters of what is morally permissible or obligatory. It would follow, then, that our duty to obey the law is our "paramount" moral obligation. Consequently, if our professional conduct conforms to the dictates of the law, then it is morally defensible. In fact, some professional organizations do appear to hold this position. Like the AAES's *Model Guide* discussed above, the code of ethics advanced by the American Academy of Ophthalmology, for example, also affirms the overriding importance of conformity to the law. It states that "the *Code* is only permitted to represent standards that fall within the law" (15). The Academy, therefore, will not require its members to act in any way that would violate the law.

The legal approach to moral behavior would appear to make sense for two immediate reasons. First, it certainly seems prudent from the viewpoint of staying out of trouble to obey the law. Second, it can be argued that the laws that govern a community reflect the considered moral judgments and sentiments of that community. This does not mean that the law is identical to morality. What it means is that there is a *private morality* that falls outside the scope of the law and a public morality that is the concern of the law. The laws in a community represent the *public morality* in a codified form.

There are objections to this kind of reasoning, however. Again from the vantagepoint of self-interest, knowledge of the law alone is no guarantee against litigation and criminal liability. The right to refuse lifesaving medical treatment, for example, has become legally enforceable in the case of autonomous and informed refusals of treatment, though it has not always been. Today competent patients can legally refuse any treatments or surgeries, including those that are lifesaving. Consequently, a competent and adult patient who is a practicing Jehovah's Witness can legally refuse a lifesaving blood transfusion. In the past, physicians would generally transfuse such a patient in spite of their refusal. Today, medical paternalism of this sort would be punishable as battery. Physicians acting purely from beneficent motives to save life during the transitional stage from the legality of such "forced" transfusions to their illegality may find themselves the object of personally damaging legal action. It is because of changes in the boundary between conduct that falls within the scope of the law and that which falls outside of it, that it may be prudent to look beyond the law to gain some idea of what is will ultimately be considered morally appropriate and morally inappropriate conduct.

There are even deeper concerns about limiting our moral obligations to the dictates of the law. We discussed some of these in relation to the criticisms of some professional codes of ethics. The foremost criticism we discussed is that the law, i.e., current laws, may be morally in error. In other words, laws are sometimes morally indefensible and require us to do things that are morally wrong. The Fugitive Slave Law of 1850, for example, required that citizens help return runaway slaves to their owners. Most would argue that this law is morally indefensible since slavery itself is morally unacceptable. The legality of abortion is a current example of a legal statute that some would consider immoral.

Another obvious concern about legal models of moral responsibility is that they are not universal. That is, what is legal, and thus moral, in one society may not be acceptable in another society where the laws differ. *Ethical Relativism* is a term used to describe the idea that the social norms that prevail in a given society are the morally appropriate norms for that society. In other words, because they are the *norms*, they are morally appropriate. The major objection to this kind of relativism, of course, is that norms are just that, *norms*. They are averages that vary from society to society; that vary within a society as it evolves; and that represent only the moral sentiments of majorities, to the frequent exclusion of minorities. Critics, then, find the legal model of moral responsibility unacceptable because of its lack of moral authority and the absence of grounding in universally accepted moral principles.

The Societal Model of Moral Responsibility

According to the societal model of moral responsibility, the prevailing social norms and attitudes of a given society determine the right-making characteristics of actions. This is sometimes referred to as *conventional morality* since it defines our moral obligations in terms of the conventions or customary rules and practices of a given society.

The problem here again is that the prevailing social norms in a given society may themselves be morally indefensible. Human sacrifice, for example, has been practiced in many cultures and has been the social norm and custom for significantly long periods of time. Many would argue, however, that it is still nevertheless wrong to sacrifice human beings despite its historical popularity or conventionality. As with the legal model of moral responsibility, the societal model can be criticized as lacking sufficient grounding in universal moral principles.

The Religious Model of Moral Responsibility

This model requires very little explanation. It states simply that right and wrong are defined in terms of prevailing religious beliefs. The difficulties here are obvious. For one, it is seldom an easy task to determine what religious beliefs prevail in a given society. Beyond this, however, is the question of whether the prevailing religious beliefs are ultimately morally justifiable. In other words, "Which is the *morally right*, or *correct*, religion?" To argue that a given religion is the correct one precisely because it is the prevailing one begs the question. We would have to know why this particular religion or these particular beliefs are morally appropriate for reasons other than that are the most popular in a given culture or community.

The Natural Law Model of Moral Responsibility

The Natural Law model of moral responsibility is far more pervasive and implicit than many people realize. Consider how often people will object to something because they claim that "It just isn't natural." For example, many people will object to the prospect of extending the human life span an additional 50, 100, or 200 years because it would not be natural; it is not the way Nature designed us. Historically, the supposed dictates of Nature and the Natural Law have been instrumental in determining moral conduct. The argument is simply that God designed Nature according to His plan and because this plan is reflected in the laws that govern Nature, it follows that we are morally obligated to obey the Laws of Nature since they are God's laws.

There are many ways to respond to the Natural Law model of moral responsibility. For example, if it is true that Nature and Nature's Laws are part of God's design and if we have some moral obligation to honor that plan, then the Natural Law (whatever that might be) would establish the parameters of morality. However, it is not obvious what Nature intends. In other words, does Nature intend that man should not fly because Nature did not give us wings or did Nature intend that we should fly by giving us the intellectual and physical means to discover and imitate flight? Consequently, if we find ourselves confused about Nature's design and just what the dictates of the Natural Law might be, then we would need to look elsewhere for moral guidance. If, furthermore, we have no reason to assume that Nature or the Natural Law has such God divined moral status or that God's design is even relevant in moral matters, then we would again have to look elsewhere for moral guidance.

Philosophical or Theoretical Models of Moral Responsibility

There are numerous models of moral responsibility that could be regarded as philosophical. There are several versions each of Utilitarianism, of Deontological Theories, of Intuitionist Theories, Libertarian Theories, Neo-Naturalism, Cognitivism, Non-Cognitivism, among others. Generally, these are philosophical models not just because they have been proposed and defended by notable philosophers but because the defense itself is philosophical. That is, proponents of these models offer as strong a justification, or defense, of their theories as possible. It is not enough, therefore, that there are codes of moral conduct that govern social interaction or that there are other competing codes of conduct. Philosophical inquiry insists on some kind of defense of these codes. It asks "What makes this particular code of ethics the 'right', or appropriate, code of ethics?"

It may be the case that there can be no ultimate, rational justification of any particular code of ethics. However, the challenge of providing such a justification stimulates a process of inquiry that probes our most cherished commitments and beliefs. It makes us think about the ultimate reasons why we believe certain acts are morally right and others morally wrong. It is this emphasis on a rational justification that we will be borrowing for the problem-solving approach suggested in this chapter.

PROBLEM-SOLVING IN ETHICS

The problem-solving model developed here involves five steps. It is a systematic approach to moral deliberation that is designed especially for groups of individuals, such as ethics committees, but can be used by individuals with some minor adjustments. It is the method recommended here because it works. The author of this chapter has seen the success of this model solving real-life ethical difficulties in various professional settings. Figure 1-1 illustrates the steps of this model. It is read from left to right.

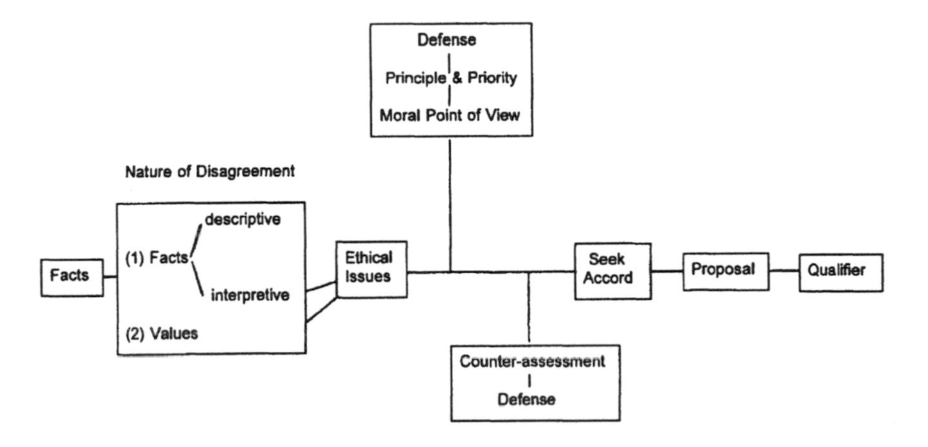

Figure 1-1. Problem-solving in ethics.

1. State the Problem

The first step in case ethics is to formulate as clearly as possible the exact nature of the ethical problem or dilemma. For example, the problem in our initial case study is that our engineer is faced with losing his job if he does not participate in the development of a product he believes would jeopardize public safety. An ethical dilemma exists precisely because there are moral reasons why our engineer should refuse to participate in the project and moral reasons why he should not. The reason we want to be clear about the nature of the problem is so that we can anticipate the kind of solution that is required. In other words, we need to be able to decide whether our engineer should participate in this project, not whether he/she should sabotage the development of the product, resign, or enter the foreign legion. We want to be able to recommend a solution or give council that is relevant to the interests at stake.

2. Get the Facts

The second step is to be clear on the facts that we have and to make sure that we gather all the relevant facts to the extent that we can. We want to be able to make an *informed* decision and to do so, we must possess and understand as best we can the relevant facts pertaining to the case. Adequate knowledge of the facts is no guarantee that a morally appropriate decision will be reached but it is highly improbable that we could even approach a morally defensible decision without adequate knowledge of the facts.

Adequate knowledge of the relevant facts of a case is important for the obvious reason that anything less would make the decision-making process both arbitrary and impertinent. But there is an added reason. Sometimes a problematic looking issue is not really a disagreement over sensitive moral issues or values but a disagreement over the *descriptive nature* of the facts. These are the "cold, hard" facts, for example, whether it's raining today in Boston or not. Disagreements involving the facts are much more easily reconciled. We simply need to get the facts straight. For example, consider a situation where a structural engineer is hired by a local newspaper to conduct an on-site inspection of a state bridge construction project. A very critical report is then published in the paper. Is there a moral problem in this case? That depends on the facts. According to the NSPE's *Code of Ethics for Engineers* "Engineers shall issue no statements, criticisms or arguments on technical matters which are inspired or paid for by interested parties, unless they have prefaced their comments by explicitly identifying the interested parties on whose behalf they are speaking, and by revealing the existence of any interest the engineers may have in the matters" (16). Assuming this rule is ethically sound, there is no moral problem if the engineer explicitly identified the interested parties and revealed any personal stakes. That the engineer did or did not do so is a factual issue, not a moral one.

The more intractable problems occur when there are disagreements over the interpretation of some important factual matters or the values that underlie conflicting moral viewpoints. For example, in medicine, only "competent" patients are recognized as having the decision-making capacity to refuse medical treatment. Consider, then, the case of a 75-year-old women who refuses a lifesaving below the knees amputation of a gangrenous limb that will poison her blood and cause her death if not removed. Yet she refuses, claiming, "I fear disfigurement far more than I fear death." Now, while it is true that adults have a legal right to refuse a lifesaving amputation, should we consider this woman a competent and capable decision-maker? If so, then we are legally required to respect her refusal of treatment; if not, then we would have grounds to dismiss her objections and perform the necessary surgery. The facts that we have to work with are that the woman is elderly and that her reason for refusing surgery is that she fears disfigurement more than she fears death. Assuming that this is all the information we have to go on, how should we interpret these facts? Is the fear she expresses the kind of reason competent persons would offer to refuse a lifesaving treatment?

Indeed, some might argue that competent decision-makers would never make a decision concerning life and death on the basis of physical disfigurement. Such reasoning might be considered *un*reasonable and an indication that the patient is not a competent decision-maker. Similarly, one might argue that refusing a lifesaving blood transfusion because it is a sin against God to "drink" blood reflects the thinking of a person who does not possess adequate decision-making capabilities.

On the other hand, others will argue that capable decision-makers could decide such matters on the basis of reasons that others might find eccentric, idiosyncratic, and perhaps peculiar. In a society that tolerates a wide range of beliefs and behaviors, persons who have idiosyncratic and eccentric beliefs and opinions will be considered no less competent than other "normal" people. The question of competence is not a scientific question. It is a question of interpretation. We are given certain facts and then we have to interpret them in terms of certain values and beliefs that we have. It is the disagreement over these moral values and deeply cherished beliefs that leads to the most vigorous moral debate.

3. Identify and Defend Competing Moral Viewpoints

When there is disagreement, or confusion, concerning the way certain facts are interpreted or over the moral values and moral principles that are at stake, we will need to examine the relative strength of the competing moral viewpoints. Hence, the 3rd step in case ethics is to assess the strengths and weaknesses of competing moral viewpoints as carefully and critically as possible. The idea here is to make the best possible case for competing moral viewpoints and defend them as thoroughly as possible.

We begin by identifying what we believe to be the most compelling reason(s) for the various courses of action available. In the opening case of this chapter, one course of action is to refuse to participate on the assigned project. The most apparent reasons for this course of action are that our engineer thinks that it would be morally wrong to participate and that the *NSPE's Code of Ethics for Engineers* appears to confirm this. But we have to ask, "Why is it wrong?" What are the moral considerations that make it wrong to participate? We must go to much greater depth defending the idea that it is wrong to do 'X', whatever 'X' might be. We need to know what makes refusing to participate the morally appropriate course of action.

In other words, it is not enough just to identify and list the reasons we think some course of action is right or wrong. This is only a partial defense. To be confident in the strength of these reasons, we must be able to justify them. We must ask what the ultimate justification is for the reasons that have been offered. It is not enough in ethics to simply appeal to a code and say "This is right because the *Code* says so." What we would need to know here is what makes the *Code* or any of its canons morally appropriate?

Consider, then, how we would defend the NSPE's first fundamental canon of ethical conduct. That canon states that engineers shall "Hold paramount the safety, health and welfare of the public." We have also noted that the NSPE Board of Ethical Review has consistently interpreted this canon to require that engineers place public safety and health above personal interest or contractual obligations. At first glance, this position may seem very defensible morally. Indeed, the unique character of moral reasoning is that it forces us to consider the interest of others as equivalent and sometimes prior to our own interests. The duties of morality are sometimes contrary to the dictates of prudence and self-interest precisely because morality generally imposes obligations to promote the interests of others over one's own interests. Truth telling, for example, is a moral duty that imposes an obligation to tell the truth even though one's personal ends might be better served by lying. This does not mean that morality does not recognize duties to oneself; it does. However, the duties to oneself are generally second to the duties we have to others.

This, however, is only one view of morality. There are others views, e.g., the egoistic view, that place self-interest above the interests of others. One might argue, for example, that our engineer has an overriding moral obligation to himself to act in a way that will further his own interests. If his interest is in keeping his job, then he may decide not to participate. If his interest is to protect the public, then he may decide to refuse to participate. Either course of action is possible from the viewpoint of self-interest. The difference is in the ultimate justification for that position. That is the point of this step in the problem-solving enterprise. We are attempting to justify our position by offering a solid defense that is grounded in compelling reasons.

4. Formulate an Opinion

Once we have considered competing positions, it becomes necessary to take a stand. Unlike purely theoretical or academic discussions, case ethics requires that we make decisions that have an immediate bearing on real-life situations. We do not have the luxury of postponing our decisions or leaving a question open and unresolved. Case ethics means coming up with a decision how to handle or resolve an ethical dilemma. Therefore, the 4th step in case ethics is to decide which of the competing moral viewpoints is the most compelling in this particular case. If a committee is making a decision, it is likely to be decided by a vote of all the committee members.

The difficult question as a practical matter, of course, is how to decide which of the competing moral viewpoints is the most compelling or most correct. When a single individual conducts this problem-solving process, the best estimation will still nevertheless be a personal estimation. This, unfortunately, reduces the problem-solving enterprise to a subjective matter of personal judgment. As with objections to a subjective model of moral responsibility generally, the same objections hold here no matter how carefully one assesses the merits of competing moral viewpoints. We are still dealing with the evaluative judgments of single persons and these may be unacceptable to others.

The committee approach to problem solving in ethics offers some distinct advantages. For one thing, it involves the judgments of several, or even many, individuals. Moreover, depending upon the composition of the committee itself, this collective judgment may be quite broad and well-informed. If the ethics committee is made up of both non-professionals and professionals from various disciplines, we might expect decisions and opinions representative of the general public at large. By contrast, a committee made up almost exclusively of members from a single profession risks the kind of bias we discussed earlier.

But let us assume that a committee is composed of persons with diverse backgrounds and interests. It would still be difficult to claim that such a committee is truly representative of the general public or that it can speak for the public at large given the diversity of public opinion. Yet, participants on committees are generally very interested in ethics and quite concerned with promoting ethical behavior in the professions. Indeed, the willingness to serve on an ethics committee says a lot about the individuals involved. This may be such a redeeming feature of committees that we are apt to take more seriously their judgments and recommendations than those of the "public," single individuals, or a single group of professionals, no matter how "expert" they may be.

Indeed, quite a number of surveys indicate that health professionals, patients, and their family members at these facilities take the opinions and recommendations of hospital ethics committees very seriously.* There is clearly something very compelling about the evaluative judgments of persons willing to donate their time and resources to ethical inquiry and discussion. Consequently, when an ethics committee can speak as a whole on ethical situations, we are inclined to listen.

The question often arises whether it is possible to get a consensus from a committee, especially one addressing such important matters as morality. There is very little in the literature addressing this question but, again, in the experience of the author, consensus is attained with remarkable frequency. This is not to say that unanimity will be reached consistently or often. But an overwhelming consensus is often reached. For example, in a random sampling of 16 cases reviewed by the NSPE's Board of Ethical Review, only one included a dissenting opinion.

Consensus is important at the practical level because it allows committees to make recommendations and offer guidance. When the committee as a whole endorses recommendations, they will be very compelling. If there is division and irreconcilable disagreement, then the opinions will be much less compelling. For example, a committee split evenly on some ethical issue, such as accepting gifts from suppliers, will be unable to offer much advice. Even a committee split 60/40 can claim something of a consensus but their judgment will not be nearly compelling as a committee achieving an overwhelming consensus of, say, nine out of ten. Of course, these numbers are arbitrary. But the reader need only ask him or herself how many of one's peers it takes to persuade and convince.

5. Qualify the Opinions or Recommendation

Because a consensus, like a verdict, is sometimes not unanimous, committees must qualify the recommendations they make by describing the level of consensus achieved. The U.S. Supreme Court does much the same thing. The qualify-

---

* See for example J La Puma, CB Stocking, CM Darling, M Siegler. Community Hospital Ethics Consultation: Evaluation and Comparison with a University Hospital Service. Am J Medicine 92(4):346-351, 1992, and J La Puma, CB Stocking, MD Silverstein. An Ethics Consultation in a Teaching Hospital: Utilization and Evaluation. J Am Med Assoc 260(6):808-811, 12 Aug 1988.

ing statement of a committee's recommendations and statement of opinion should include the voting distribution and any dissenting opinions.

APPLYING THE COMMITTEE APPROACH TO SPECIFIC CASES

Now let's apply this method to the case at the opening of the chapter. The **first step** is to formulate the problem briefly and to the point. We decided in a previous section that the problem in this case is that our engineer will lose his job if he does not participate in a project that he believes would jeopardize public safety.

The **second step** is to collect all relevant facts. We are given very few facts to start with, so we should think about the things that we would need to know, given the circumstances in this case. For example, what kind of contract, if any, does our engineer have with his employer? Is he allowed to refuse a job assignment for reasons of conscience? If so, then he needs to remind his employer that he cannot be fired for refusing to participate in this project. If not, his concerns are quite real.

Another question we might ask is whether this job is worth keeping. Are other equal or better job opportunities available? If so, the risk to our engineer's financial security is not nearly as great as it would be if this were the only job in town. Another question worth asking is whether our engineer is correct in his assessment of the use of radar detectors in general and the possible threat to public safety. In other words, are his concerns well founded? If there was no threat to public safety, then there would be no ethical conflict between self-interest, the employer's interests, and the public interest. If there is the possibility of such a threat, then the ethical conflict is very real.

The **third step** requires that we examine competing moral considerations or viewpoints. Reference to our list of moral principles and duties will help clarify the principles, duties, and values and stake. Refer back to the end of the section on the ethical principles to see how they are used to help describe the ethical conflict in this case.

This third step requires substantially more than simply describing the ethical conflict in terms of competing *prima facie* moral principles, however. It also requires that we make the best possible case for each of the conflicting moral viewpoints by offering the strongest justification or defense that we can. Consider, then, what the strongest defense might be for the overriding importance of our duties of nonmaleficence and beneficence in this particular case.

According to the principles of nonmaleficence and beneficence, we have moral duties not to inflict harms and moral duties to promote the good of others and to prevent and remove harms. The essence of both these principles is expressed in the first canon of the NSPE's *Code of Ethics for Engineers*. As noted earlier, it tells us that engineers must "Hold paramount the safety, health, and welfare of the public." This canon may be interpreted to express at least three action-guides: first, that engineers should not inflict harm; second, they should promote the good of others; and, third, these obligations are of overriding importance "in the performance of their professional duties." If we accept that marketing the "Stealth Radar Detector" will indeed jeopardize public safety, then it would follow that participating in the marketing of such a device is an act that presents the risk of harm to others. The possibility of "inflicting" such harm is enjoined by the principle of nonmaleficence. Moreover, we can promote the good, viz. the physical safety, health, and welfare, of others, by doing what we can to make sure that the public cannot get its hands on a device whose purported sole purpose is to facilitate speeding. This would accord with our duties of beneficence. We must still ask, however, why these duties are overriding.

One possible defense for the overriding importance of the principles of nonmaleficence and beneficence in this case is that the NSPE's *Code* makes them so. However, this appeal to the authority of the *Code* is only a partial defense. What is it at the core of these principles that makes them so important in this case? Why would promoting public safety be more important for an engineer than, say, his or her own financial security, the financial prosperity of the engineer's employer, or freedom of choice in the marketplace. I would invite the reader to consider this question in depth. One plausible response is that engineers have been awarded the public's trust as builders of the infrastructure that enables the public to engage in cooperative social enterprise. Engineers, therefore, have special role-specific duties to keep the public trust by making public safety paramount in the practice of their profession. In this sense, engineers may have obligations of fidelity to keep their end of an implied social contract that obliges them to maintain the overriding importance of these principles. Licensing requirements are one way this contract is made more tangible.

Now consider a defense of the position that makes duties to respect autonomy overriding. What would that defense look like? We might argue that the public to whom the marketing of this device is directed is composed of autonomous, self-governing, and self-determining persons whose freedom to choose in the marketplace ought to be respected. If they purchase this device and endanger public safety using it to evade detection while speeding, then they are the parties responsible for violating the duty not to inflict harm on others, not the engineer who designed the device.

Ultimately we are probably going to have to address all these issues of safety versus freedom, personal responsibility, autonomy, etc. But eventually and, very often, soon, we will have to decide. This is the **fourth step** in our problem-solving approach. At some point, discussion of competing viewpoints must end and we are faced with making a decision. The decision, of course, should address the initial problem. In this case, we need to know whether our engineer is morally required to refuse to participate in the design of the "Stealth Radar Detector." In the committee setting, the decision and recommendation will be reached through parliamentary procedures and a vote. We cannot know what that ultimate decision would be in a case like this. And, indeed, much of the outcome will depend on the strength of each participant's convictions going into the discussion and the job we do "selling" each of the competing viewpoints.

Once the deciding has been done, the decisions and recommendations of a committee must be qualified, as instructed in the **fifth step** of this model.

CONCLUDING THOUGHTS: THE GOOD AND BAD OF AN ETHICS COMMITTEE APPROACH TO MORALITY

An ethics committee approach to morality is interesting both politically and philosophically. First, the opinions of ethics committees deserve our attention because they represent the considered judgments of persons interested in and caring enough to involve themselves in the problem-solving process of case ethics. Philosophically, this problem-solving enterprise is interesting because it points toward a democratization of moral values, rules, and principles. Some will find this a positive feature and some a negative feature. On the one hand, the ethics committee approach offers a democratic approach to establishing moral rules and principles by providing a conduit for our moral judgments and opinions.

On the negative side, democratic processes of any kind pose the risk of a possible "tyranny by the majority" (unless certain protections are built-in to safeguard the interests of real or potential minorities). Critics will also argue that "morality by majority" usurps the moral authority of some possible "higher" authorities, such as the Bible, the church, the state, or certain alleged "experts." For many people, these higher authorities are the ultimate basis that justifies a given moral viewpoint. According to this view, a democratic process of determining morality not only risks being in error, but is fundamentally an objectionable approach to moral decision-making.

A democratic process to determine morality also means that there are no absolute moral principles and rules or an absolute prioritization of them. Nothing save the democratic process itself is clearly, universally, and without exception right or wrong. Indeed, something becomes right or wrong because we decide democratically that it is right or wrong. This will certainly rub some people the wrong way. But, ethics committees are an increasingly useful and utilized resource in this society. They provide a forum for ethical inquiry and decision-making. They draw upon the expertise and sensitivities of many dedicated persons. They offer objectivity in situations often fraught with emotion and personal interest. And they reduce the burden of decision-making by spreading responsibility among many decision-makers. Moreover, the ethics committee process does not necessarily preclude the role of religion or religious tenets. These interests will often be at the core of defenses offered for particular moral viewpoints. Thus, the problem-solving model we have discussed in this chapter is offered as a practical and compelling method for addressing ethical dilemmas and reaching solutions for those dilemmas as they arise in the engineering profession.

STUDY QUESTIONS

1.1 This chapter discussed the dilemma of the engineer who is asked to design and develop a radar detection device. He has misgivings about developing the device yet believes that he has an obligation to work for the best interests of his employer. What should the engineer do? Should he refuse to undertake the assignment thereby possibly jeopardizing his job? Or should he do as his employer asks? Why or why not? Justify your conclusions in light of the *Code of Ethics* of the National Society of Professional Engineers.

1.2 Consider the following case from "You Be the Judge," a regular column appearing in *Engineering Times*, the newsletter of the National Society of Professional Engineers:

N. A. Quandry, a principal in ABC Engineering, submits a proposal to a local municipality to be considered as the consultant for the research and analysis of a former dump site that is being considered for reclamation as a wetland. In a meeting, the municipality indicates the possibility that there could be hazardous and toxic wastes in the dump. Upon being awarded the contract, Quandry is informed that he must sign a confidentiality clause that precludes him from disclosing any results or information concerning the project without the city's written permission. Quandry signs the contract and the clause.

Preliminary tests of the surface soils on the site are inconclusive but reveal a possibility that very high contaminant levels of hazardous and toxic waste could, over time, become exposed at the surface and wash into a river that flows immediately adjacent to the site. The city is considering plans to build a childrens' park, recreation and picnic area, bike/jogging trail, and parkway near the reclaimed areas, and the river is used for drinking water intake. Upon receiving the initial data, the city terminates the contract and, as its reasons for not continuing, cites the political fallout of revealing the findings and the economics of having to clean up the property. Quandry responds that the city has a responsibility to the public to proceed to remediation, but the city refuses and reminds Quandry of the confidentiality clause and the legal consequences of going public with the confidential information.

Is Quandry bound by the NSPE *Code of Ethics* to inform the appropriate regulatory agencies of his findings and the potential dangers to the public health and the environment? Did Quandry behave ethically in signing the confidentiality clause restricting him from revealing information concerning dangers to the public health and the environment after being informed that the site could contain hazardous and toxic wastes?

1.3 What makes "right" actions right? That is, why are some actions morally correct and others morally incorrect?

1.4 Do you think that the ethical requirements of engineers should be more or less demanding than the ethical requirements of other professions? Why or why not?

1.5 Why is engineering as a profession more demanding ethically than other professions?

1.6 How is *compliance* with a company's policies and rules different from morality?

1.7 How can compliance with company policies and rules help engineers avoid ethical difficulties?

1.8 How is compliance with a professional code of ethics different from morality?

1.9 How rigidly do you think professionals should adhere to their respective professional codes of ethics?

1.10 What are some advantages that a company and its employees might gain from having its own ethics committee?

1.11 What are some disadvantages of company-formed ethics committees? In other words, what are some concerns that people might have of company ethics committees?

1.12 Consider who you would turn to first when faced with an ethical dilemma. Why did you make this choice?

REFERENCES

1. 42 U.S.C. 1395, 1396 (1990).
2. JS Mill. On Liberty. New York: W. W. Norton & Co., 1975, p. 12.
3. TL Beauchamp, JF Childress. Principles of Biomedical Ethics. 2nd ed. New York: Oxford University Press, 1983, p. 148.
4. WK Frankena. Ethics. 2nd ed. Englewood Cliffs, NJ: Prentice-Hall, 1973, p. 47.
5. J Rawls. A Theory of Justice. Cambridge, MA: Harvard University Press, 1971, pp. 60-62.
6. National Society of Professional Engineers. Code of Ethics for Engineers. Section I, Canon 1. Alexandria, VA: July 1996 (see Appendix).
7. NSPE Board of Ethical Review Case No. 65-12, Alexandria, VA: 1965.
8. NSPE Board of Ethical Review Case No. 82-5, Alexandria, VA: 1982.
9. NSPE Board of Ethical Review Case No. 88-6, Alexandria, VA: 1988.
10. NSPE Code of Ethics for Engineers. Preamble. Alexandria, VA: July 1996.
11. American Association of Engineering Societies. Model Guide for Professional Conduct. Washington, DC: December 13, 1984 (see Appendix).
12. NSPE Code of Ethics for Engineers. Section III Rule 3a. Alexandria, VA: July 1996.
13. NSPE Code of Ethics for Engineers. Section III, Rule 4. Alexandria, VA: July 1996.
14. NSPE Code of Ethics for Engineers. Section II, Rule 1e. Alexandria, VA: July 1996.

15. American Academy of Ophthalmology. Ethics in Ophthalmology: A Practical Guide. San Francisco: AAO, 1986.
16. NSPE Code of Ethics for Engineers. Section II, Rule 3c. Alexandria, VA: July 1996.

2

Defining Engineering Ethics

BLACK AND WHITE, OR GRAY?

In the May 1998 newsletter of the North Carolina State Board of Registration for Professional Engineers and Land Surveyors, the board chairman, R. Larry Greene, RLS, said:

> *When I was a young man, acting in an ethical manner seemed to be a straightforward proposition. An act was either right or wrong, black or white, morally correct or [not].... As an older, perhaps wiser, man I still try to guide my life by those principles, but I have learned over time that recognizing right, wrong, black and white is not always a simple proposition.... Every situation we find ourselves in is different from any other we have ever experienced. As each new scenario is added to a job situation, the ethical mix changes and the boundary between ethical and unethical conduct, perhaps already gray in nature, shifts.*

On the other side of the world, in the June 1997 issue of *The Building Economist*, the journal of the Australian Institute of Quantity Surveyors, Tony Fendt, managing director of Project and Retail Pty. Ltd. said:

> *We work in an uncertain world; an increasingly global economy where the competition for employment increases by the day, and the pool of opportunity decreases by the day whether due to population increase, advancement in technology, downsizing, increased per capita production, etc.*
>
> *The question is are we justified in stretching the truth, or even concealing the truth, in order to gain or retain favour with a client?*

In order to answer this question, we need to examine our moral roots. Ethics is primarily concerned with what we ought to do when what is right and what is advantageous and profitable conflict with one another.

Mr. Fendt goes on to quote Cicero's letter "On Duties" written to Cicero's errant son, Marcus:

Let us regard this as settled: what is morally wrong can never be advantageous, even when it enables you to make some gain that you believe to be to your advantage. The mere act that believing that some wrongful course of action constitutes an advantage is pernicious."

Chapter 1 defined ethics as "...what we ought to do and how we ought to behave from a moral viewpoint, as opposed to an economic, religious, political or prudential viewpoint."

These various comments illustrate the clear relationship of ethics to morality and also suggest that the definition of ethics is strongly dependent upon current cultural and social viewpoints. What is obviously ethical and moral in one context may be quite the opposite in another. In engineering, the distinction between black and white is often even less clear. Engineers operate in a multinational, multicultural business environment in which what is considered moral and ethical often varies from one location to another. In some cultures it is legal and totally acceptable morally for substantial gifts to be exchanged between those desiring to do business in that area and those seeking to have work done. In Western cultures, this is considered to be bribery and is illegal; in other cultures it is normal, acceptable business practice. Later in this book several case studies will be presented addressing this issue.

Ethics clearly is not black or white—it is many shades of gray depending upon the given situation. Nevertheless, engineers do have guidance in determining what ethical standards they should apply to their life and work—the codes of ethics of their professional societies. These codes define for the engineer what is acceptable and what is not. They define what engineering ethics is, and what it is not. Every engineer and engineering student should become thoroughly familiar with the code of ethics of his or her disciplinary engineering society and the *Code* of the National Society of Professional Engineers. They form the foundation for ethical practice and, not coincidentally, generally are written in whole or in part into the laws and regulations governing the practice of engineering in the United States.

THE *CODE OF ETHICS* OF THE NATIONAL SOCIETY OF PROFESSIONAL ENGINEERS

The *Code of Ethics* of the National Society of Professional Engineers is the preeminent document defining the ethical responsibilities and obligations of engineers practicing in the United States. The complete *Code* appears, with the permission of NSPE, in the appendix of this book along with the codes of several other engineering and engineering-related organizations.

The NSPE *Code* begins with six very brief fundamental canons which form the foundation for the much more detailed "Rules of Practice" and "Professional Obligations" which make up the great majority of the *Code*. The fundamental canons and their preface read as follows:

Preamble

Engineering is an important and learned profession. As members of this profession, engineers are expected to exhibit the highest standards of honesty and integrity. Engineering has a direct and vital impact on the quality of life for all people. Accordingly, the services provided by engineers require honesty, impartiality, fairness and equity and must be dedicated to the protection of the public health, safety and welfare. Engineers must perform under a standard of professional behavior that requires adherence to the highest principles of ethical conduct.

Fundamental Canons

Engineers, in the fulfillment of their professional duties, shall:

1. *Hold paramount the safety, health and welfare of the public.*
2. *Perform services only in areas of their competence.*
3. *Issue public statements only in an objective and truthful manner.*
4. *Act for each employer or client as faithful agents or trustees.*
5. *Avoid deceptive acts.*
6. *Conduct themselves honorably, responsibly, ethically, and lawfully so as to enhance the honor, reputation, and usefulness of the profession.*

This preamble and the six very brief canons, in and of themselves, probably constitute an adequate code of ethics for any profession. Look at them and consider replacing *engineering* and *engineers* with *medicine* and *medical doctors* or *the legal profession* and *lawyers*. Do that and you would have an excellent set of ethical guidelines for those two professions, or for virtually any other profession.

COMMON SENSE VS. THE FUNDAMENTAL CANONS

Is there an engineer anywhere who doesn't relate to Dilbert®, Scott Adams' comic strip about an engineer spending his days working in a cramped cubicle, surrounded by incompetent co-workers, and dealing with the infamous "pointy-haired boss"? Probably not. Too many things that happen to Dilbert seem to strike home to us engineers (and in fact many of the ideas for the strip are contributed by engineers to Scott Adams by e-mail).

A Dilbert strip published on May 30, 1995 struck home because Dilbert and the pointy-haired boss were discussing ethics. The gist of the strip was that the boss asked if Dilbert had taken a required ethics training course. He told the boss that he hadn't and suggested that if the boss said that he did, money would be saved which the boss could spend for himself. The boss responded that luckily, he hadn't taken the training either. Dilbert replied that ethics "... is mostly common sense anyway."

This author is not advocating that anyone should falsify business records or use funds for an improper purpose, but Dilbert is correct that ethics is mostly common sense. If you think that the *NSPE Code of Ethics for Engineers* is a legalistic, incomprehensible document, read it. After doing so, you will agree with Dilbert's statement.

Consider the "Fundamental Canons" of the *Code* which form its basis: They begin by saying engineers shall "hold paramount the safety, health and welfare of the public." Who could disagree with that?

Next they say engineers shall "perform services only in areas of their competence." Only a fool would do otherwise. This canon is clearly common sense.

The third canon says engineers shall "issue public statements only in an objective and truthful manner." This is common sense too, if for no other reason than the reaction of the media to improper statements.

Next we are to "act for each employer or client as faithful agents of trustees." Why would we do otherwise?

Fifth, we are to "avoid deceptive acts." Clearly that makes sense.

And finally, engineers are told to "conduct themselves honorably, responsibly, ethically, and lawfully so as to enhance the honor, reputation and usefulness of the profession." Again, this is pure common sense for any profession.

The *Code* goes on to amplify upon the six Fundamental Canons with specific Rules of Practice and a list of Professional Obligations, all of which also make common sense.

The message here is that if you wonder if any contemplated action of yours is ethical or not, it probably isn't. Your common sense should give you the correct answer.

Read the *Code* in the Appendix and you undoubtedly will agree that what it says is little more than a statement of common sense practices for engineers. If you want a separate copy for reference, check it out on the NSPE Internet web page at http://www.nspe.org/eh1-code.htm and click on the "print" command to make a hard copy. While you are browsing the Internet, check out Dilbert. The address is http://www.unitedmedia.com/comics/dilbert/. And, if you have any good ideas for the comic strip, on engineering ethics or otherwise, Scott Adams would also appreciate hearing from you. You can contact him at scottadams@aol.com.

THE SCOUT LAW

Many engineers are leaders in the Boy Scouts of America and come face to face each week at Scout meetings with a short, twelve-point "code of ethics," the Scout Law, which states that:

> *A Scout is trustworthy, loyal, helpful, friendly, courteous, kind, obedient, cheerful, thrifty, brave, clean, and reverent.*

That is a lot shorter than the *NSPE Code of Ethics for Engineers* but does the NSPE *Code* say much more that the Scout Law? I really don't think so. The *Code* is very detailed and specific and the Scout Law is very general but says just as much, if not more.

Can people *trust* in your work as a Professional Engineer?

Are you *loyal* to your clients, employers, the public, and government?

How *helpful* are you? Do you do your best to assist your clients and employers, your peers, and your subordinates?

Are you *courteous* and *kind* to all? Courtesy goes a long way in promoting good business relationships. An unkind word or action can irreparably damage business relationships.

Are you *obedient* to the reasonable and proper desires and needs of your clients, employers, and subordinates?

Are you *thrifty* in your designs by making them of suitable quality to do the job which is necessary without unneeded or undesired "gold plating"?

What about *bravery*? Are you willing to stand up for your beliefs and to speak out against those things which you know are improper or unethical?

Do you make sure that your office and your project sites are kept *clean* and free from impediments and debris? It only makes sense and avoids many safety hazards.

Reverent? Are all of your actions guided by what your faith tells you is proper?

An article in the March/April 1998 *Scouting Magazine*, the magazine Scout leaders receive, is entitled "Run the Twelve," referring to the twelve points of the Scout Law. The author, Harold R. Limke, is an engineer in Evergreen, Colorado and is also a Scoutmaster. When he runs into an ethical dilemma with his Scouts, he tells them to "run the twelve." In other words, he says that whatever they do should be guided by those twelve principles. He says, "You can test any action before you take it by running the twelve. Compare your actions against the twelve points of the Scout law. Is my action trustworthy? Is it loyal? Is it helpful? And so on." Mr. Limke is quite correct. The next time that you wonder if something is ethical, try running the twelve.

STUDY QUESTIONS

Consider the following case from the column "You Be the Judge" which appears regularly in *Engineering Times*, the newsletter of the National Society of Professional Engineers:

2.1 Tess R. Groundwater, PE, an environmental engineer, is retained by a major industrial owner to examine certain lands adjacent to an abandoned industrial facility formerly owned and operated by the owner. The owner's attorney, Rich N. Slick, requests that as a condition of the agreement Groundwater sign a secrecy provision whereby she would agree not to disclose any data, findings, conclusions, or other information relating to her examination of the owner's land to any other party unless ordered by a court. Groundwater signs the secrecy provision.

What do you think? Was it ethical for Groundwater to sign a secrecy provision agreeing not to disclose and data unless ordered by the court?

2.2 N. Byast, PE, is retained to provide both design and construction-phase services. After construction starts, a dispute arises between the engineer's client and the general contractor concerning the acceptability of a concrete pour. Byast seeks to remain impartial in the dispute, citing a provision in his contract with the client that makes the engineer the initial interpreter of the contract requirements and the judge of whether the work is acceptable. The client and the contractor ask Byast to review the dispute. Following his review, Byast agrees with the contractor's position, noting that the client had approved certain changes in the work and the contractor had complied with those changes. The client accepts Byast's interpretation, but also criticizes him, claiming that because of his ethical duty of loyalty to the client, Byast should have found in the client's favor.

Did Byast owe an ethical duty to the client find in the client's favor?

2.3 Edgetown has retained Lota Ventures, PE, an engineer in private practice, as town engineer. The Edgetown Planning Board is reviewing for approval a project being proposed by ABC Development Enterprises. Ventures is also being retained by ABC Development Enterprises on a separate project being constructed in Nearwood, a town in another part of the state. The Nearwood project is unrelated to the project under consideration by Edgetown. Ventures is expected to offer her views in the capacity of town engineer to the Edgetown Planning Board as to the feasibility of ABC's Edgetown project.

Would it be ethical for Ventures to develop and report her views on the feasibility of ABC's Edgetown project to the Edgetown Planning Board?

2.4 The Old Eastbranch Department of Public Works retains Johnson and Witherspoon Engineering to prepare plans and specifications for a water extension project. Ed Minister, PE, chief engineer of the department, instructs Johnson and Witherspoon to submit its plans without showing the name of the firm on the cover sheets, but permits it to show the firm name on the individual working drawings. The department also makes it a policy not to show the name of the design firm in the advertisements for construction bids; in fact, the advertisement states "plans and specifications as prepared by the Old Eastbranch Department of Public Works."

Was it ethical for Minister to adopt and implement a policy that prohibited the identification of the design firms on the cover sheets for plans and specifications? Did he act ethically in authorizing the statement in the advertisement?

3

Ethical Technologies in Engineering, Construction, and Project Management*

THE CHERNOBYL NUCLEAR DISASTER

April 26, 1986 was a significant day in the Northern Hemisphere. Yet few will remember why. It was the day that Reactor No. 4 at the Chernobyl Nuclear Plant experienced a melt down, an explosion, and subsequent massive nuclear fallout. Much has been written about the causes of the melt down. A number of writers who have spoken to witnesses and seen public records attribute it to a group of electrical engineers who were "experimenting" with this potentially highly unstable reactor, by shutting it down to test run-on of the turbine.

Because the instability was well known, many safeguards—electrical and mechanical—were built into the reactor to prevent people from shutting it down. The Chernobyl nuclear station is depicted before the melt down in Figures 3-1 through 3-4. These pictures are publicity photographs, which were given to visitors at Chernobyl. From these pictures, the plant has every appearance of being a well-designed, modern engineering facility. Nevertheless, despite the safety systems built into the design of the plant, No. 4 reactor was shut down and *each* safeguard was overridden by the electrical engineers. The consequences, as shown in Figures 3-5 and 3-6, are well known. The real question here is not why the reactor exploded but the ethical question of what ethical and moral codes were the engineers working to that enabled them to knowingly override warnings of disaster in pursuit of their experiment.

----

* By Dr. Peter J. Rutland, Waiariki Institute of Technology, Rotarua, New Zealand

Figure 3-1. The Chernobyl Atomic Station before the disaster.

Figure 3-2. Chernobyl Atomic Station control room.

Another critical point is that the technology of the nuclear plant had a significant impact on the result of unethical behavior. Had this sort of behavior happened in the nineteenth century, the resulting effect would have been nothing like the scale of the Chernobyl disaster. The early twentieth century concepts of ethics were generally of the kind that encouraged ethical attitudes and behavior in business, but if they did not exist it was not considered to be devastating.

Figure 3-3. Interior of the Chernobyl Atomic Station.

Figure 3-4. Another interior view of the Chernobyl Atomic Station before the disaster.

Figure 3-5. Picture of Chernobyl Reactor Building No. 4 after the disaster. (Source: Russian Research Center "Kurchatov Institute".)

Figure 3-6. The "Elephants foot", the once liquefied radioactive core of the Chernobyl reactor. (Source: Russian Research Center "Kurchatov Institute".)

This approach is totally unsatisfactory today as is evidenced by the chemical plant disaster in Bhopal, India, AIDS-tainted blood transfusions in France, and mad cow disease, since the scale of technologies is different not just in level but in kind. Technological systems are in place, as are human systems, such that the injection of unethical behavior at the higher levels can produce unmitigated disasters. Figures 3-7 and 3-8 show the temporary solution to the Chernobyl problem, construction of a massive concrete "sarcophagus" to encase the highly radioactive remains of Reactor No. 4, a temporary solution at best. A sarcophagus may successfully entomb an Egyptian king for centuries but not a decaying mountain of nuclear waste.

Figure 3-7. Construction of the "sarcophagus" to entomb the Chernobyl reactor. (Source: Russian Research Center "Kurchatov Institute".)

The large-scale, interconnected, technologically advanced systems of the twenty-first century present a potentially huge ethical problem. The Internet has already demonstrated that. Although very intelligent people operate these systems, they are capable of causing disasters if there are no ethical guidelines to be followed. Guidelines must exist and must be followed.

Many of the science fiction stories intelligent people devour on film are based on the unethical villain capturing a worldwide system which he or she can then turn to advantage by wreaking havoc on the poor unsuspecting good people of the world. Fortunately, Captain Kirk or Batman or 007 always arrives to save the day. Such superheroes don't exist in reality, but the potential villains do.

Figure 3-8. The finished "sarcophagus" in winter 1986. (Source: Russian Research Center "Kurchatov Institute".)

The degree of "smartness" or intelligence in the operations of the large-scale systems is increasing geometrically. Not because each individual is getting much smarter, but because computer technology is advancing so quickly that robotic, voice-responsive software systems can control the systems when given instructions by people. Agreed, these people are likely to be very intelligent also, and one could well argue that this is itself a safeguard against disaster. But Chernobyl, the Exxon Valdez, Bhopal, and other disasters do not support that theory. Large systems operated by super-intelligent software packages designed and manipulated by the smartest people can be a recipe for disaster if the smartest people are not ethical.

ETHICAL CHALLENGES IN ENGINEERING, CONSTRUCTION, AND PROJECT MANAGEMENT

Undoubtedly the myth of amoral business persists and indeed a common observation is that the phrase *business ethics* is an oxymoron. But such a view is dependent on the subjugation of personal interests that insist on business people acting under the guidance of a moral philosophy that is often contrary to business itself.

The nature of projects themselves present many ethical concerns (1). First, in the project initiation and feasibility stage, there are concerns about such things as falsification of estimates, invalid requests for proposals that are really only an effort to obtain project ideas, and concerns about the ethical responsibilities of external consultants.

When the project progresses to the planning and organization stage, many more areas of ethical concern arise. Bid rigging is a frequent problem area and involves divulging of confidential information to some bidders in an effort to influence the amounts of the bids or to give some bidders an unfair advantage over other bidders. In other cases, the bidding process itself is little more than a sham because the bids are "wired"—the winner is predetermined.

"Low balling" is another major problem with contractors attempting to "buy" the project by bidding low in the expectation of recovering any costs via subsequent change orders, contract renegotiations, or simply by cutting corners.

Bribery is another big problem area, particularly in international project work. This is discussed in some detail in Chapter 13.

Another ethical dilemma exists with firms declaring their capability to perform while not truly being able to do so.

Also, in this stage the problem of falsified estimates, both of cost and schedule, may arise, as may problems of discrimination in hiring.

In the implementation and execution phase many additional concerns may arise including padding of expense accounts, using substandard materials, compromising health and safety standards, and withholding information from clients, owners, or superiors, etc. The list goes on and on.

Finally at project closing, problem areas such as failing to honor commitments to project personnel, failure to recognize or admit project failure, and sloughing off to protect one's position can occur.

Ethical dilemmas are common for engineers, project managers, and construction managers. Among those identified in discussions with a number of project professionals are:

- being offered gifts from contractors or vendors
- pressuring to alter status reports with backdated signatures or faded documents to mask reality of project status
- compromising quality
- falsifying reports of charges for time and expenses
- lowering the quality of communication with co-workers and management and clients
- abusing power and openness and transparency of information

ETHICAL CODES AND PERSONAL VALUES

Many organizations do have ethical codes in written form, which provide a useful framework for the ethics in decision making. Six such codes appear in the appendix of this book: the codes of AAES, NSPE, the Project Management Institute (PMI), the Association for the Advancement of Cost Engineering (AACE International), the American Society of Civil Engineers (ASCE), and the Institute of Electrical and Electronic Engineers (IEEE). These codes are typical and like most others are concerned with relationships between co-workers, dealings with outside interests, conflicts of interest, and dealings with clients and with the community. The NSPE *Code of Ethics for Engineers* and the AAES *Model Guide for Professional Conduct* have been discussed in Chapters 1 and 2. The PMI *Code of Ethics for the Project Management Profession* deals with the standards of personal professional conduct, operations at work, relations with employers and clients, and responsibilities to the community. AACE International, like PMI, is heavily concerned with project management and construction management, in addition to engineering in general. AACE's *Canon of Ethics* is similar in content to the NSPE *Code*, as is the ASCE *Code of Ethics*. The IEEE *Code* is somewhat briefer than the NSPE code and is more similar to that of AAES.

Despite this unanimity of professional thinking, discussions with engineers, project managers, and construction managers indicate that codes of ethics are not seen as being the most essential means of guaranteeing ethical conduct. Most people spoken to indicated their own personal ethical standards and values provided more guidance than any formal documentation. It is encouraging to know that the majority of people place personal values very highly on the list of satisfiers in their work even to the extent that if their personal values were compromised, they would consider leaving.

THE ETHICAL FRAMEWORKS

There are many definitions of ethics, but some ethical theories have been classed by Cullen, Victor, and Stevens (2) into:

Egoism, which is optimizing one's own interest, i.e., the highest gain is to pursue one's own welfare;

Utilitarianism or optimizing the interest of one's self and significant others, and;

Deontology or designing to do what is right regardless of the actions, the specific outcome, and whose interests are affected by it.

Take, for example, a bribe offered to an individual to ensure that a contractor gains a contract. If these theories were tested against reality, then the egoist would take the bribe or even ask for a bigger bribe, the utilitarian could well refuse the bribe, and the deontologist would probably report this to the law or ombudsman.

A more comprehensive approach is by Vesilind (3), who classifies ethical frameworks according to cost/benefit judgements or contractually based obligations, duties or commitments; the degree to which they are directed towards means (i.e., the act or the end results); and the degree to which they are directed towards serving self or others. This tri-dimensional approach means that, in the cost/benefit based model, concepts of hedonism and egoism are directed towards self, utilitarianism is directed towards the end, and altruism is basically focused on others. In a contract-based model the deontological approaches are focused on the means where religious services focus on others and social contracts are based on ends.

This particular typology and that of Cullen, Victor, and Stevens are very useful in describing motives and the theories behind why people behave in certain ethical manners. They do not necessarily provide guidance on values, and therefore do not help engineers, project managers, and construction managers know how they should act in given situations. What they can do is allow project professionals to assess how they have reacted in terms of the theory. Then, of course, recriminations or remorse may well set in.

ETHICAL TECHNOLOGIES

Each of the challenges identified above will be larger in the twenty-first century than they ever were before because of the power of technology to allow them to proceed. It is thus paramount that the potential impact of ethical unfitness is assessed. Questions that need to be asked are:

1. What is the extent of interconnectedness of this activity?
2. What level of smartness is attached to it?
3. What level of ethical fitness exists in the operations?

A tri-dimensional approach such as this allows each project or element of a project to be assessed in ethical terms, just as they are in financial and chronological terms. For example, a project has been established to develop a network of satellites which can be used to monitor telephone calls and identify specific words or phrases.

A project such as this will obviously have high interconnectedness and high smartness and, therefore, requires a high level of ethical fitness if it is not to be used to the disadvantage of many people.

SUMMARY

This chapter has highlighted some of the ethical issues faced with advancing technology, infrastructure, and engineering and project management into the twenty-first century. The concept of inter-connectivity, smartness, and ethical fitness are addressed and a tri-dimensional model proposed which allows ethical decision making to be placed within it to determine the amount of ethical fitness required.

The model must not be misused by indicating that lower levels of interconnectedness and smartness require no ethical fitness. Some level of ethical fitness will almost always be needed. Members of the engineering, project management, and construction management professions are encouraged to actively pursue the ethics of technology into the twenty-first century in order to avoid monumental disasters.

STUDY QUESTION

3.1 The Chernobyl disaster involved one reactor of four at the Chernobyl nuclear station. The remaining reactors continue to be operated as do other reactors of similar design in other countries. It is widely acknowledged that the overall design of this type of reactor is inherently dangerous and that human failings, such as those which caused the disaster at Chernobyl, could cause a similar such disaster in the future.

The Chernobyl disaster on April 26, 1986 exposed thousands of humans to dangerous levels of radiation and caused radioactive dust to rain down across much of Europe. Despite these facts, the "Chernobyl type" reactors continue to be operated, in great measure because no adequate alternate source of electric power is readily available in the affected countries.

Conversely, the Three Mile Island nuclear failure in the United States proved to be a non-event. The reactor failed and partially melted down but the radiation was contained within the plant. No member of the general public was harmed nor was the public ever in danger from the incident. The light water nuclear reactor technology used in the United States was proven by the Three Mile Island incident to be far superior to that at Chernobyl. The primary reason is that U.S. designed reactors are completely contained within a protective structure to prevent escape of any significant amount of radiation in the unlikely event of a melt down or explosion. Nevertheless, the combined public and political reaction to the Chernobyl and Three Mile Island accidents has resulted in an almost total cessation of nuclear plant construction in the United States. Electric power reliance has instead been shifted to construction of more fossil fuel (coal, oil, and natural gas) burning power plants which are known to contribute to environmental degradation, air pollution, acid rain, depletion of the ozone layer, and global warming.

Given these facts, discuss the ethical responsibilities of engineers involved in continued operation of "Chernobyl type" reactors in Eastern Europe and the ethical responsibilities of American engineers involved in the design, construction, and operation of light water nuclear reactors.

REFERENCES

1. DJ Robb, Ethics in Project Management Issues Practice and Motive, PM Network, December 1996.

2. JB Cullen, B Victor, C Stevens, An Ethical Weather Report: Assessing the Organisation's Ethical Climate Organisational Dynamics. Autumn, 50-62, 1989.
3. PA Vesilind, Views on Teaching Ethics and Morals, Journal of Professional Issues in Engineering Educational Practice, 117: 88-95, April 1991.

4

Continuing Professional Development in Engineering

PROFESSIONAL ENGINEERING LICENSING

Because of the rapid rate of advancement of technology, engineering graduates are literally obsolete in their level of technical knowledge almost immediately upon graduation from college. Consider something as common and widely used as the computer and the Internet. Changes are so rapid that what is an adequate level of knowledge today most certainly will be inadequate in a year or less. The rate of change is phenomenal and, in order to keep up to date, any engineer must make a concerted effort at maintaining his or her skills and thus must be deeply concerned with continuing education and continuing professional development.

Historically, professional engineering licenses were lifetime approval to practice engineering provided that renewal fees were paid in a timely manner as required by the various licensing and regulatory boards and agencies. No evidence of continued professional competence was required, and any engineer could renew a license to practice simply by payment of a fee—even though that engineer might be technologically obsolete. That situation is rapidly changing and more and more state licensing authorities are requiring engineers to prove that they have made some concerted effort to maintain their skills. Iowa, Alabama, and West Virginia were the first states to implement such requirements, and the trend is such that all states will inevitably implement some sort of continuing education or continuing professional development requirement for license renewal as a Professional Engineer.

To promote uniformity in these requirements among the states and to help insure continued reciprocal recognition of licenses from state to state, the National Council of Examiners for Engineering and Surveying (NCEES) promulgated a "model rule" for states considering adoption of mandatory continuing professional development (CPD) requirements.

The NCEES model rule, which has generally been accepted by those states that have adopted mandatory CPD requirements, is based upon the concept of a Professional Development Hour (PDH). NCEES defines the PDH as one nominal contact hour (generally a typical 50-minute lecture or class hour) of instruction or presentation. By comparison, a Continuing Education Unit (CEU), a nationally recognized unit of measure for continuing education, is equal to 10 hours of instruction or training. Thus one CEU is equivalent to 10 PDHs.

The NCEES "model rule" recommends the following credits for approved PDH activities:

1. 1 college or unit semester hour....................................45 PDH
2. 1 college or unit quarter hour ....30 PDH
3. 1 Continuing Education Unit (CEU) ....10 PDH
4. 1 hour of professional development in course work, seminars, or professional or technical presentations made at meetings, conventions, or conferences............................1 PDH
5. For teaching, apply a multiple of 2 (first time only); teaching credit does not apply to full-time faculty.
6. Each published paper, article, or book ....10 PDH
7. Active participation in professional and technical societies (each organization).....................................2 PDH
8. Each patent...10 PDH

The model rule requires licensees to acquire 15 PDHs per year for relicensing. The rule also permits a maximum of 15 PDH credits to be carried over into the subsequent renewal period. However, not all states accept this latter rule, and some limit the carryover to less than 15 PDH.

Obviously, 15 hours of effort per year does not insure continuing professional competence, but the requirement does illustrate the concern of the regulatory bodies about the need for engineers to make every possible effort to continue their education as a lifelong effort. The various codes of ethics that are included in the Appendix of this book stress, in one way or another, that engineers shall perform services only in their areas of competence. With the rapid rate of advances in technology, it is imperative that engineers make every effort to continually upgrade and improve their skills. To do otherwise would violate one of the fundamental canons of the profession.

ENGINEERING CERTIFICATION PROGRAMS

Licensing of professional engineers is generally restricted by the states to the specific disciplines taught in the universities and colleges. However, by the very nature of the profession, engineers often become specialists in very narrow engineering specialties rather than in the broader discipline in which they were originally trained. Through experience and continuing education, it is not uncommon for an engineer to become quite knowledgeable in an engineering specialty that is quite remote from the engineer's original field of training and licensure. The various codes of ethics and licensing laws recognize this fact by permitting professional engineers to practice in any area of the profession in which they have competence. The problem with this concept is how competence can be defined. Is it sufficient for an engineer to think or say that he or she is competent in a particular narrow specialty or must some other method be used to assess competence?

Early Certification Programs

The preceding question was the catalyst for development, beginning in the 1950s, of a plethora of programs purporting to measure competence of engineers in a multitude of specialty disciplines. Some of these programs, such as those in quality control, environmental engineering, and cost engineering, were highly regarded. Others were little more than a sham whereby persons could be issued some sort of certificate or piece of paper avowing competence in a specialty simply by paying some person or organization a fee. There are many certification programs in existence, some of which issue certificates for a lifetime without examination or proof of competence, while others impose very rigorous standards, far more rigorous than anything required or anticipated by any registration law. It is necessary to distinguish between the bona fide certification programs and those that are without substance.

Cost engineering certification was one of the earlier programs which began in 1976 when the American Association of Cost Engineers (now the Association for the Advancement of Cost Engineering) implemented a program in response to member demand for a credential to recognize competence in the field. Eight years later, in 1984, the Project Management Institute implemented a certification program for project management professionals. These programs have grown rapidly and similar programs are now in effect in other parts of the world. The certification trend, however, is not limited to cost engineering and project management—it is proliferating in virtually every engineering specialty, including project management, which is very closely related to cost engineering.

The obvious questions to ask about certification are, "Why is it needed?" "Are not the academic degrees and professional licensure enough in the way of credentials?" and "What purposes do additional credentials serve?" There are strong advocates and strong opponents of specialty certification (and of PDH programs for license renewal). The opponents see it as an unnecessary waste of time and money and an obstacle to many competent professionals. The proponents feel that it is necessary because of the high degree of specialization that now exists in the engineering profession and because licenses and degrees, by their very nature, are measures of broad general training and knowledge, not specialized expertise in a narrow technical area.

Historical Background

In 1988 a major conference was held in Atlanta, Georgia for the purpose of sorting out some of the issues about engineering certification. Participants included representatives of 23 national and international engineering and technical societies. In two days the participants thrashed out their diverse viewpoints on the advantages and disadvantages of specialty certification and whether or not it indeed had any significant value vis-à-vis professional licensure and academic degrees.

A joint consensus statement that summed up the conclusion of the debates was adopted at the close of the conference. It reads:

WHEREAS, the National Conference on Engineering Specialty Certification is the first profession-wide meeting to discuss this subject, and

WHEREAS, the Conference has explored a wide range of issues and viewpoints on specialty certification for engineers;

NOW THEREFORE, it is the sense of the conference that the individuals attending this conference believe that specialty certification for engineers, properly conceived and administered, can be valuable to both the public and the engineering profession.

Considering the strong opinions, pro and con, that the participants possessed going into the conference, this conclusion was indeed of major importance.

The conferees also agreed to a statement authorizing "...the Conference Planning Committee to continue as a liaison committee for any further development of a national organization for specialty certification."

The outgrowth of this activity was the formation on April 24, 1990 of the Council of Engineering and Scientific Specialty Boards (CESB). The bylaws of CESB define its goals and objectives as the improvement of engineering practice for the public benefit through:

A. *the establishment of professional criteria and monitoring of qualifications used to recognize special capabilities in engineering and related fields of practice through specialty certification;*

B. *the encouragement of continuing professional development as a condition for continued specialty certification; and*
C. *the encouragement of ethical practice as a condition for continued specialty certification.*

The environmental engineering profession pioneered specialty certification in engineering in 1955, embracing the same principles found in medical specialties—formal training at the college level, a state license to practice engineering, a prescribed amount of practice experience, a review of qualifications by peers, and examinations.

There are now a great many other engineering and engineering-related organizations operating specialty certification programs. The formation of CESB in 1990 was another step in the evolution of these efforts. The purpose of CESB is to instill some uniformity in engineering and technical certification programs and to identify those credible programs on which the public can rely.

The accreditation decisions are based on a comprehensive assessment of each program's operations against prescribed guidelines adopted by CESB. These guidelines specify certain responsibilities to individuals, consumers, and the public, which the program must fulfill together with the methods it used to certify and recertify. A fundamental facet of the CESB guidelines is a requirement for periodic renewal of certification certificates, which is based upon continuing professional development activities in much the same manner as the NCEES "model rule" applies to licensure. CESB accreditation however requires a substantially greater number of PDH credits than are required by the NCEES "model rule."

It is also important to understand that CESB itself does not certify individuals. It accredits certification programs. The actual certification is done by an accredited technical society.

Defining Certification

Part of the misconceptions surrounding certification lie in the incorrect application in society of the words used in the credentialing field, i.e., *licensing, registration, certification, and accreditation.* Engineers, at least in the United States, use the term *registration* interchangeably with *licensing*, which contributes to the confusion since the two words have sharply different meanings to the public. A license is authorization granted by a government to perform a function or service, e.g., a driving license or an engineering license. The root of *licensing* lies in the police powers of government to protect the health, safety, and welfare of the public.

Registration is listing with and by some body. It can be a governmental or nongovernmental entity that does the registration, e.g., registering as a voter. It grants no authority nor does it address qualifications.

Certification is a voluntary act that in some organized fashion measures an individual's qualifications to perform a specialized function. Because it is voluntary, it conveys no authority or privilege, i.e., one does not need to possess the certificate to perform a function or service, albeit custom or market forces may require it. Certification exists today in many, many professions and trades.

Accreditation is like certification, except that it applies to institutions and programs, not individuals. A familiar example is the accreditation of schools.

An excellent definition of engineering certification is that of Dean Peter Chiarulli of the Illinois Institute of Technology some years ago. He said:

Engineering certification is a process through which an individual obtains peer recognition of his competency to engage in engineering or technical activities in a particular engineering or technical specialty field.

That is all it is. Peer recognition that you are qualified to be engaged in engineering work. Peer recognition—this is quite different from licensing or mandatory continuing professional development activity for license renewal.

Engineering certification is totally analogous to the medical profession wherein a physician is licensed to practice in his or her state or country and then is certified through board examinations administered by medical societies, not licensing boards, in one or more specialties—e.g., family practice, cardiology, surgery. The distinction is significant. Professional peers, not licensing boards, establish standards for continuing professional practice that must be met to maintain certification.

We need to ask who is the better judge of competency in highly specialized fields of practice: one's peers, or a regulatory board that is broadly based and that does not necessarily have expertise in the particular specialty in question. Certification programs attempt to answer that question.

Roots of Engineering Credentials

Modern engineering is a far cry from its roots in designing the engines of war for medieval battles. The first specialization to occur was to separate civil engineering from military engineering. Today *Who's Who in Engineering* lists more than 100 different U.S. engineering organizations, each with its own special interest. Many hundreds of similar organizations exist around the world.

Wyoming was the first U.S. state, in 1907, to exercise its constitutional powers to regulate the practice of engineering. Between then and 1947 all states passed registration laws for engineers, and by 1970 all 50 states and five legal jurisdictions of the U.S. had laws that regulated engineering practice in some way. However, unlike medicine and law, these regulations contained exemptions allowing persons to practice engineering without a license. The net result of these exemptions is that nearly two-thirds of the two million practicing engineers in the United States are not licensed.

The lack of a universal requirement for licensing mandated by government, as well as practical differences of opinion between licensing boards and certain practice groups, led many engineering specialties to eschew registration and pursue other forms of self-regulation, be it association membership or certification. These differences and a host of underlying factors, personalities, and power struggles are another primary contributor to the ongoing debate regarding specialty certification in engineering.

The purpose of formally recognizing or accrediting certification programs is to provide the engineering profession and the general public with assurance that certification programs do meet a set of rigorous standards. This is completely analogous to the accreditation of academic degree programs and serves the same function. Specialty certification in engineering is a reality, and the process of accreditation is the quality control mechanism for certification programs.

Unique to the engineering psyche is an inborn resistance to regulation. The attributes that make good engineers and scientists—creativity, independent minds, and fact orientation—exacerbate this tendency. Further, it is human nature to fear failure, and accordingly, to avoid scrutiny and examination.

However, in a complex society embracing millions, the wishes of the individual must give way to the needs of society to protect itself from those who would prey on its vulnerabilities. Recognizing and accommodating these needs is the essence of professionalism. Grappling with those conflicting motives is the essence of the ongoing debate in the engineering profession regarding specialty certification. Our technical societies have the mission to meet the needs of society, and this is the real underlying purpose of certification. As in medicine, the objective is to assess competence and continuing professional development in engineering specialties by those most qualified to do so—one's peers—not a government agency or regulatory board.

Specialty certification programs should exist to address the objective of public service. If properly conceived and administered, they build quality and competitiveness. This is CESB's mission in identifying and recognizing those programs that meet these altruistic goals.

STUDY QUESTIONS

4.1 An engineer holds an accredited degree in civil engineering and is licensed as a Professional Engineer. Through experience he has become a specialist in cost engineering (cost estimating, cost control, and planning of major construction projects) and practices almost exclusively in this specialty area. He considers himself to be an expert in this field. He has become aware that the Association for the Advancement of Cost Engineering offers a specialty certification program in cost engineering and that this program has been accredited by CESB. Should he consider sitting for the cost engineering examination? Why or why not?

4.2 An engineer works for an industrial firm that produces manufactured products for the consumer market. Because of the industrial exemption clause in his state laws, he is not required to be licensed as a professional engineer and therefore has made no effort to become licensed, considering the license to be unnecessary. The consumer products his company produces include kitchen appliances that have sharp edges, rotating parts, electric shock potential, and other potential hazards. His company suggests, but does not require, that all engineers in the firm apply for registration. The company agrees to pay the required fees for refresher courses and license application. The engineer worries that he may have considerable difficulty in passing the examination, particularly the portion on fundamentals of engineering which covers much of the content of basic science and engineering courses which he took many years earlier. He fears that he will have forgotten much of that information and that it will require a great deal of study on his own time to refresh his knowledge, time which he is unwilling to give since, in any event, he is not required to be licensed. He therefore decides not to pursue licensure. What are the ethical considerations of his decision? Has he acted ethically? Why or why not?

5

Professional Engineering Licensing and Registration Laws

WHAT IS PROFESSIONAL LICENSING OR REGISTRATION?

As was briefly mentioned in Chapter 4, licensing is authorization by a government to perform a function or service. Thus, a license as a professional engineer (PE) is government authorization to practice engineering. Professional engineering licensing is often referred to as registration and licensed PEs are commonly called Registered Professional Engineers.

Unlike the medical and legal professions, however, engineers are not always required to be licensed in order to practice. The various state laws generally provide exemptions from licensing for engineers who are engaged in certain types of work. The exemptions most often apply to engineers who work for industrial firms. Thus the exemption clauses are commonly referred to as the *industrial exemption.*

Because most engineers work for industrial firms and generally are not required to be licensed, the great majority of these engineers do not seek licensure. This is a big error, one that often causes great problems for these engineers later in their careers when they change jobs, enter the consulting field, or become involved in litigation.

The easiest time to begin the licensing process is during the final year of undergraduate study in engineering, and all engineering students should do exactly that. The licensing procedure requires that the candidate sit for two examinations before becoming licensed. The first is an examination on the fundamentals of engineering and science. After several years of practical on-the-job experience, the candidate sits for the second examination that covers the principles and practice of engineering.

Once the candidate successfully writes the fundamentals examination, he or she is designated by the state as an Intern Engineer (IE) or, in some jurisdictions, as an Engineer-in-Training (EIT or ET). This is analogous to a medical doctor who serves an internship before being admitted to practice in hospitals without direct supervision by other physicians. The IE or EIT must serve a period of years gaining experience under the supervision of a professional engineer before being permitted to sit for the final licensing examination.

Why should the engineering student sit for the IE or EIT examination before graduation? The most compelling reason is that while still a student the candidate has the greatest possible chance of being successful on the examination. The examination is very comprehensive and covers the full breadth of undergraduate technical courses. Therefore, by sitting for examination during the final undergraduate year, the course material is still fresh in the candidate's mind and the examination is generally fairly easy to write. However, if the candidate waits several years before sitting for the fundamentals examination, memory fades. Much of what was studied at the university may not have been used in actual practice and thus may have been forgotten. The result is that a very high percentage of those who attempt to write the fundamentals examination after graduation are not successful and are thus not able to become licensed. In some cases this means an abrupt end to an employment offer, a planned consulting practice, or even termination from current employment. It is easy to rationalize that the license will never be needed but that is a very risky decision. No one knows what the future will bring and the wise student therefore takes the first possible opportunity to write the fundamentals examination.

Another consideration is that there is a clear trend toward eliminating exemptions from registration or to severely reducing the scope of existing exemptions. Several states have effectively eliminated the industrial exemption and others will inevitably follow suit. Engineers who are currently exempt from registration may suddenly find that they must become registered just to retain their current job, let alone to change positions.

WHAT ARE THE REQUIREMENTS FOR LICENSING/REGISTRATION?

State requirements for licensing as a professional engineer vary but, in most cases, first require that the candidate hold a four-year engineering degree which has been accredited by the Accreditation Board for Engineering and Technology (ABET) or that the candidate hold an equivalent degree from a foreign college or university.

The second requirement is that the candidate has successfully passed the fundamentals examination.

Most jurisdictions then require four or more years of experience working under the supervision of a licensed professional engineer, followed by the successful completion of the principles and practice examination in the candidate's disciplinary engineering field.

In addition, the candidate must provide favorable character references and must also provide references from professional engineers who are familiar with the candidate's engineering work, preferably PEs who have been supervisors of this work. These persons preferably should have been in *responsible charge* of the candidate's work. What does that mean? General Statue 89C of the State of North Carolina and Chapter 30, Article 13, of the West Virginia Code both defines it as *" 'Responsible Charge'... direct control and personal supervision...of engineering work...."* Similar definitions are included in the laws of the other states.

J. Albert Bass, Jr., PE, a member of the North Carolina Board of Registration for Professional Engineers and Land Surveyors explained responsible charge in this way (1):

"Responsible Charge" is the relationship an engineer...is required to have with another person who is preparing drawings or other engineering work for the professional, and which will be sealed by the professional.

In other words, the professional references submitted by the candidate should preferably come from a licensed professional engineer who bore the ultimate responsibility approval of the candidate's work. The laws of most states say that the candidate must have *"...a specific record of...four years or more of progressive experience on engineering projects of a grade and character which indicate...that the applicant may be competent to practice engineering."* The wording in the laws varies but this statement is the gist of what most of these state laws say.

Once the engineer has become licensed, it may be wise to apply to the National Council of Examiners for Engineering and Surveying (NCEES) Records Verification Program. A license is generally valid only in the jurisdiction which issued it, and in order to engage in engineering practice in another state or jurisdiction a license will most probably be required from that state or jurisdiction. Because engineers often must work in several states and frequently are transferred or seek employment in another state, the licensing procedure can become very cumbersome and onerous. The NCEES Records Verification Program offers a less cumbersome way for obtaining licensure in multiple jurisdictions. Once the first license is obtained, the PE can apply to NCEES for inclusion in the Records Verification Program. NCEES will verify and hold the PE's file containing college transcripts, licensure information, professional references, and employment verifications. When the PE seeks licensure in a new jurisdiction, a copy of the NCEES record can be transmitted to the other licensure authorities. This transmittal generally substantially reduced the amount of paperwork required in order to obtain the additional license or licenses.

The NCEES record does not guarantee licensure but it is a recognized and reliable source of information for the various registration boards. The NCEES record generally simplifies and expedites licensure via comity (reciprocity) in the great majority of jurisdictions. For information on the program, contact NCEES at

PO Box 1686, Clemson, SC 29633-1686 or see *http://www.ncees.org* on the Internet.

ETHICS AND THE LAW

In addition to the advantages of licensure or registration that were cited earlier, the PE has special protection under the law by virtue of being licensed. The licensed engineer has been legally recognized as being competent to practice engineering. This can be very significant if the engineer's work is ever questioned in a court of law or if the engineer appears as an expert witness in court proceedings. The licensed engineer is presumed to be technically competent unless it can be proven otherwise. The unlicensed engineer, on the other hand, is not considered to be competent until this can be clearly demonstrated in court. This is truly a significant distinction and, in itself, is sufficient reason for any engineer to seek to become licensed.

However, the granting of a license to practice engineering carries with it a special obligation to practice in an ethical manner. The NCEES has developed a set of model rules of professional conduct as guidance to state engineering licensing boards in laws and regulations. The NCEES rules are based upon the NSPE *Code* provisions. State licensing boards generally have statutory authority to promulgate their own set of rules, which are binding upon every person holding an engineering license permitting them to offer or perform engineering services in the state. Most states and territories incorporate or paraphrase the NCEES model rules or portions of the NSPE *Code* in their regulations governing the practice of engineering. For example, The North Carolina Administrative Rules governing the practice of engineering and land surveying in that state incorporate virtually all of the Rules of Practice which appear in the NSPE *Code* (2). The North Carolina rules also specifically require professional engineers having *"...knowledge or reason to believe that another person or firm may be in violation of any of these provisions...[to]...present such information to the board in writing and ...[to]...cooperate with the board in furnishing such further information or assistance as may be required by the board."* Thus the North Carolina licensed engineer has a legal obligation to practice in accord with the *Code of Ethics* **and** to report others who fail to do so. It is not sufficient for the licensed engineer to practice ethically. It is also legally incumbent upon that engineer to assure that others also practice in an ethical manner and to report those who violate ethical standards or who are suspected of doing so.

Most state and territorial licensing and registration boards have adopted similar rules of practice, generally based upon the NSPE *Code*.

In some states, ethical requirements have been written into law rather than being promulgated as rules of practice by the licensing board. Consider the following excerpt from the Engineer, Land Surveyor, and Geologist Registration Law (Act 367, Section 4(g)) of the Commonwealth of Pennsylvania (3):

It shall be considered unprofessional and inconsistent with honorable and dignified bearing for any professional engineer...

(1) *To act for his client or employer in professional matters otherwise than as a faithful agent or trustee, or to accept any remuneration other than his stated recompense for services rendered.*

(2) *To attempt to injure falsely or maliciously, directly or indirectly, the professional reputation, prospects, or business of anyone.*

(3) *To attempt to supplant another engineer...after definite steps have been taken toward his employment.*

(4) *To compete with another engineer...for employment by the use of unethical practices.*

(5) *To review the work of another engineer...for the same client, except with the knowledge of such engineer... or unless the connection of such engineer... with the work has terminated.*

(6) *To attempt to obtain or render technical services or assistance without fair and just compensation commensurate with the services rendered: Provided, however, the donation of such services to a civic, charitable, religious or eleemosynary organization shall not be deemed a violation.*

(7) *To advertise in self-laudatory language, or in any other manner, derogatory to the dignity of the profession.*

(8) *To attempt to practice in any field of engineering... in which the registrant is not proficient.*

(9) *To use or permit use of his professional seal on work over which he was not in responsible charge.*

(10) *To aid or abet any person in the practice of engineering...not in accordance with the provisions of this act or prior laws.*

This section of the Pennsylvania law specifically empowers the Pennsylvania Board to discipline any registrant who violates this code of ethics. The Pennsylvania Board has also promulgated additional regulations governing the ethical practice of engineering in that state.

John R. Speed, PE, Executive Director of the Texas State Board of Registration for Professional Engineers said (4):

...As engineers, ethics should be considered a part of every action. More specifically, ethics is the way we act. Do we achieve our goals in a manner that is trustworthy? What is the impact of our actions on others? Do we act based on instinct, or do we act upon that which we know? Do we present our actions to others in an honest and complete manner?

I propose that the most fundamental elements of ethical behavior center around three things: objectivity, honesty and trustworthiness. Our board rules reflect these critical facets of ethical behavior, and our enforcement cases document the failure of engineers to work within them.

Ethical practice of engineering is the law in North Carolina, in Pennsylvania, in Texas and in every other state and territorial jurisdiction of the United States. There are no exceptions. The message should be clear—engineers, whether licensed as professional engineers or not, not only have a moral obligation to practice ethically, they are legally required to do so.

RESOLUTION OF ETHICAL ISSUES AND COMPLAINTS

As is stated earlier, professional engineers have a legal obligation to practice in accord with the ethical requirements of their state or territorial registration board and to report others who fail to do so. That requirement is clear. If the engineer knows of a violation, it must be reported to the registration board. Failure to report a violation is, in itself, a violation and can lead to disciplinary action and possible revocation or suspension of the license of the engineer who fails to make a report.

But what if the issue is not clear? What if it is uncertain if a violation has indeed occurred? What if all the engineer has to go on is hearsay or rumor and not direct knowledge? Should the suspicions be reported? If the engineer is not sure, is there any avenue to clarify the issue?

The answer is "Yes." Each state and territorial engineering society that is affiliated with NSPE has an ethics committee that can assist in resolving such concerns. The engineer can seek the assistance of the state society ethics committee directly or via the local NSPE chapter. Cases can also be referred directly to the NSPE Board of Ethical Review although the state route is preferred because it is more expeditious. The specific circumstances will then be evaluated by professional peers and, if they determine that a breach of the rules and regulations governing the practice of engineering may have occurred, they will refer the matter to the state registration board for further investigation. Throughout this process, the complaint will be held confidential, but if it is referred to the state board it is possible that the engineer who made the original report of suspected wrongdoing might be called to testify.

The state ethics committees can also assist engineers in deciding if an action they or their company is considering is acceptable. At times the issues are not clear and the state committees are willing to assist engineers in deciding upon an appropriate course of action.

In resolving ethical concerns such as these, three types of issues must be examined: *factual issues*, *conceptual issues*, and *moral issues*. Factual issues involve the truth or falsehood of the claim. Conceptual issues relate to the meaning or scope of a term or a concept. Moral issues deal with the relevance or application of moral principles to the issue at hand.

Consider the following situation:

An engineer leaves Company A to work for Company B. While at Company A, the engineer signed a confidentiality agreement with Company A concerning proprietary information. A new project at Company B on which he is asked to work involves emission of Compound X. Emissions of Compound X are not currently regulated but, from his experience with Company A, the engineer knows a method to change the process, thereby eliminating the emissions of Compound X. Further, the engineer's knowledge would not harm the competitive position of Company A. Under these circumstances, should the engineer approach Company B with the proposed change?

The ethical issues which must be addressed here are:

1. Factual issue—Is Compound X really hazardous?
2. Conceptual issue—What is "proprietary information"?
3. Moral issue—Even though it is not regulated, can Compound X cause adverse health effects?

These are the issues that must be clarified prior to determining if the engineer's anticipated actions are acceptable. The state ethics committees are a valuable resource in resolving issues such as these. Then, if the issue involves actions which have already occurred and which appear to be inappropriate, the question can be referred to the registration board for appropriate action.

As a cautionary note, the committee procedure should not be used to avoid reporting known violations or suspected violations for which strong supporting evidence is available. In such cases, the concern must be reported to the registration board without delay. It is wise, however, to report the issue to the state committee simultaneously, as it may be able to assist the board and/or the complainant in resolving the complaint.

STUDY QUESTIONS

5.1 Consider the following case from the "You Be the Judge" column of *Engineering Times*, the newsletter of the National Society of Professional Engineers.

Honus Owen, a professional engineer in mechanical systems, is a sole practitioner in a consulting firm in State X and has a business card indicating that he is a PE. However, Owen's license comes from State Y, where the bulk of Owen's projects are constructed. Cli Yent contacts Owen to design a project for construction in State X. After Owen completes the work, Yent learns that Owen is licensed only in State Y and that Owen has not obtained any authority to perform services in State X. Yent must now have another PE either redesign the project or carefully review Owen's work before sealing it. As a result, Yent will incur additional expenses and delay in the construction of his project.

Was it ethical for Owen to imply that he was registered in State X? Was it ethical for him to design the project for construction in State X? Why or why not?

5.2 The following case appeared in the *Engineering Times* "You Be the Judge" column. IJK, Inc., a firm that helps companies locate professional engineers who perform building inspection services, contacts Bill Dhing, PE, to do such work. IJK is also involved in assisting employees who relocate in selling and buying their residences. Typically, IJK makes contact with a client, takes an order for a job, and passes the order on to the professional engineer available in the geographic area. In his work for IJK, Dhing performs the services, prepares the report, and submits it along with an invoice for half of what he would normally charge another client for the same services. Dhing later learns that not only has IJK occasionally modified reports without consulting him but has invoiced clients at twice the amount charged by Dhing. IJK has no exclusive contractual or business relationship with Dhing and possesses no engineering experience.

Is it ethical for Dhing to continue his association with IJK after learning that the referral firm has a history of changing reports? Can Dhing ethically continue this association knowing that IJK invoices clients for his services at more than what Dhing charges IJK?

5.3 Assuming that Bill Dhing in Problem 5.2 was registered in the Commonwealth of Pennsylvania, did he act ethically in charging IJK only half of his normal fee? Did he in any other way potentially violate the Pennsylvania rules? Explain the reasons for your conclusions in detail.

5.4 Should Dhing take any specific actions with regard to the fact that IJK modified his reports? If so, what actions should he take?

5.5 If Dhing was licensed in North Carolina and IJK did business there, is Dhing required to report IJK to the authorities? If so, on what basis?

REFERENCES

1. North Carolina Board of Registration for Engineers and Land Surveyors. News Bulletin No. 32. November 1988.
2. Rules of Professional Conduct. North Carolina Administrative Code, Title 21, Chapter 56, Section.0701.
3. Pennsylvania State Board of Registration of Professional Engineers, Land Surveyors, and Geologists. Newsletter. Fall 1995.
4. JR Speed. Ethics and the Professional Engineer. Texas Professional Engineer. November/December 1997.

6

The "Engineers Creed" and the "Order of the Engineer"

THE ENGINEERS CREED

Physicians have a creed, the Hippocratic Oath, which says:

You do solemnly swear, each by whatever he or she holds most sacred

That you will be loyal to the Profession of Medicine and just and generous to its members

That you will lead your lives and practice your art in uprightness and honor

That into whatsoever house you shall enter, it shall be for the good of the sick to the utmost of your power, your holding yourselves far aloof from wrong, from corruption, from the tempting of others to vice

That you will exercise your art solely for the cure of your patients, and will give no drug, perform no operation, for a criminal purpose, even if solicited, far less suggest it

That whatsoever you shall see or hear of the lives of men or women which is not fitting to be spoken, you will keep inviolably secret

These things do you swear. Let each bow the head in sign of acquiescence

And now, if you will be true to this, your oath, may prosperity and good repute be ever yours; the opposite, if you shall prove yourselves forsworn.

Physicians take this oath when they enter the practice of their profession and often refer to it. This oath is, in effect, a synopsis of the physicians' code of ethics. It is brief but it says all that needs to be said.

Engineers have a similar oath, the *Engineers Creed*, developed in response to a desire for a short statement of philosophy of service and similar to the Hippocratic Oath for medical practitioners. Just as physicians often place a framed copy of the Hippocratic Oath in their offices as a constant reminder of their ethical obligations, engineers frequently display a copy of the Engineers Creed in their places of work.

Figure 6-1 shows the Engineers Creed. A parchment copy, suitable for framing, may be obtained from the National Society of Professional Engineers, 1420 King Street, Alexandria, VA 22314, at a nominal cost.

Look at the Engineers Creed. Read what it says and compare it to the NSPE *Code of Ethics for Engineers* and the other codes that appear in the Appendix. It is much shorter than any of the codes but in substance it says just as much. It is a pledge given by engineers—a pledge that reinforces the engineers' commitment to practice in accord with the highest ethical standards. In less than 100 words, the Creed says more than many people could say in a book longer than this one. Many meetings of professional engineers begin, or conclude, with the engineers reciting the Creed in unison. It is a fitting way for any engineering society to express in a corporate way the ethical basis of the engineering profession.

THE ORDER OF THE ENGINEER

That Iron Ring and That Failed Bridge

That iron ring! What is it? You may have seen an engineer wearing a plain ring on his or her little finger and wondered what it was. It looks like a simple wedding band and in a manner of speaking it is. It is the symbol of an engineer who is "married" to the profession. It is the outward sign that the engineer is a member of the Order of the Engineer.

What is the Order of the Engineer? Why do engineers wear the ring? How and where did the custom start? It all goes back to a wrought iron bridge that was being built in 1907 over the St. Lawrence River at Quebec City in Canada. At the

Engineers Creed

As a Professional Engineer, I dedicate my professional knowledge and skills to the advancement and betterment of human welfare.

I pledge:

To give the utmost of performance;

To participate in none but honest enterprise;

To live and work according to the laws of man and the highest standards of professional conduct;

To place service before profit, the honor and standing of the profession before personal advantage, and the public welfare above all other considerations.

In humility and with need for Divine Guidance, I make this pledge.

Figure 6-1. The Engineers Creed. (Source: National Society of Professional Engineers, Alexandria, Virginia.)

time this bridge was to be the world's largest single cantilevered span. Unfortunately the bridge proved to be underdesigned, and it fell into the St. Lawrence, killing 76 construction workers. The engineers had deviated from accepted procedures, tried to save some time and effort, undercalculated the weight of the bridge, and underdesigned key structural members.

In the early 1920s H. E. T. Haultain, a mining engineering professor at the University of Toronto, was seeking a method of improving the image of the engineering profession in the public's eye. He envisioned a ceremony similar to that of physicians when they take the Hippocratic Oath. He wanted a formal ritual for inducting young engineers into their profession and asked Rudyard Kipling to write the words for such a ceremony. Kipling developed a ceremony called "The Ritual of the Calling of an Engineer" in which each engineer accepts an obligation to practice in a professional and competent manner and to "honourably guard" the reputation of the profession. Upon accepting the obligation, the engineers are presented with a ring to wear on the small finger of the working hand.

The first ring ceremony was held at the University of Toronto in 1925 with rings claimed to be made of "hammered iron," the same material as the failed St. Lawrence bridge. While it is not known with certainty whether or not the first rings were actually made from the wreckage of that catastrophic engineering failure, whatever the source of the material, the rings were a permanent reminder of an engineer's responsibility. The circular shape of the ring itself is to symbolize the continuity of the profession, and it has become the virtual universal symbol of the profession in Canada where more than 99% of all engineers wear the ring.

In Ohio in 1953 correspondence began between members of the Ohio Society of Professional Engineers and the Canadian Wardens, who are responsible for the Canadian ceremony, with a view to possibly extending the Canadian ceremony to the United States. Due to copyright and other conflicting factors, extension of the ritual to this country was not possible.

Figure 6-2. The logo of the Order of the Engineer.

In 1966 a group of Ohio engineers proposed a similar ceremony for U.S. engineers using a stainless steel ring. Dean Burl Bush of the College of Engineering at Cleveland State University and his students seized upon the idea, designed a ring ceremony, and on June 4, 1970 inducted 170 engineering seniors and engineering faculty members into the "Order of the Engineer." Each participant signed a creed expressing dedication to the ethics and professionalism of engineering. Each of those inductees received a plain stainless steel ring placed on the small finger of his or her working hand. Thus began the custom of the engineer's ring in the United States, a custom which has now spread to every state and which has become an integral part of NSPE and State Society meetings, as well as those of other engineering societies.

The ring is the sign of the engineer's profession and symbolizes the engineer's obligation and promise to practice the profession in an ethical and competent manner. If you don't already wear the ring, take the first opportunity you have to join the Order of the Engineer and take on this visible symbol of professionalism. Unlike other organizations, the Order has no meetings other than the ceremony of induction, there are no dues, and the only monetary cost is a small induction fee to cover the cost of the ring and the ceremony. If you are an engineer or a graduating engineering student, when given the opportunity to participate in the ceremony and ritual, you should do so. The ring becomes the outward sign that you are an engineer who has publicly taken a pledge to practice in an ethical and professional manner.

The Obligation of the Engineer

The Order of the Engineer ceremony includes the public acceptance of an obligation, both orally and by signing a statement bearing the obligation. That obligation, in part, reads:

> *I am an engineer. In my profession I take deep pride. To it I owe solemn obligations...As an engineer, I pledge to practice integrity and fair dealing, tolerance and respect; and to uphold devotion to the standards and dignity of my profession...I shall participate in none but honest enterprises. When needed, my skill and knowledge shall be given without reservation for the public good. In the performance of duty and in fidelity to my profession, I shall give the utmost.*

The ceremony concludes, in part, with the statement:

> *This is your ring. In times of anxiety, look on it and take courage. In times of honor, regard it with humility. Wear it proudly and with distinction for you are an engineer.*

The ring indeed does tell the world that you are an engineer. It looks much like a simple wedding band except for the fact that it is worn on the little finger of the working hand. It isn't obvious but those who see it will ask about it if they don't know. A good reply is, "It is a wedding band. It symbolizes my marriage to the engineering profession." Indeed it does.

Many persons who know what it signifies recognize the ring. When it recognized, its meaning is clear. While crossing an international border, a customs agent asked me, "What kind of an engineer are you?" In surprise, I answered, "A chemical engineer but, how did you know I was an engineer?" She replied simply, "From your ring."

THE FIFTY-NINE STORY CRISIS

The *New Yorker* magazine (1) told a story about William J. LeMessurier, a renowned structural engineer, who demonstrated the utmost in observing the principles of the Engineers Creed, and who did so at great personal sacrifice.

Mr. LeMessurier had designed the structure of the Citicorp Tower in New York, at the time the seventh tallest building in the world. The structure weighed 25,000 tons and was widely acclaimed for its technical excellence and beauty. LeMessurier was 52 years old at the time and received one of the highest honors of the engineering profession, election to the National Academy of Engineering.

The building was unique in that it was supported by four massive, nine-story-high columns placed at the center of the four walls, rather than the conventional placement in the four corners of the building. This design was chosen so that the building would stand above a church, almost appearing suspended in the air. It was truly a breathtaking design.

After the building was occupied, LeMessurier received a telephone call in June of 1978 from an engineering student who questioned the safety of the design. At first, LeMessurier dismissed the comment as not being well founded. After all he was a renowned structural engineer and the student was, well, just a student.

However that call got LeMessurier to thinking about how the building had been designed. In addition to the columns, the building had been designed with wind braces for protection in severe winds. The strength of the braces had been determined to be in full compliance with New York City's building code which was based upon forces created by perpendicular winds of hurricane force, winds of the magnitude occasionally encountered in New York.

Even though LeMessurier honestly believed that the student was incorrect, he decided to go back and recalculate the forces, which the building would be subjected to in the event that the winds hit at 45 degrees instead of a perpendicular direction. He was surprised to discover that a quartering wing could increase the strain by forty percent.

That discovery might not have concerned him but he had recently learned that the braces had been joined with bolted joints, instead of the welded joints originally specified by LeMessurier. This had been done to save cost since welded joints are much more expensive than bolted joints. Welded joints are extremely strong, generally stronger than the metal itself. Bolted joints, however, are generally considered to be technically acceptable. Therefore the building had been erected with bolted joints. LeMessurier's was unaware that the change had been made. When he had learned of the change, LeMessurier initially was not concerned. He felt that the choice of bolted joints was a technically acceptable alternative to welding.

However, the new knowledge that quartering winds could increase the strain on the building braces raised great concerns, and further analysis showed that the 40 percent increase in tension caused by quartering winds would produce a 160 percent increase in strain on the bolts. The design safety factor would not cover such an increase, and the building could indeed collapse in hurricane force winds. No one involved in the design and construction of the building had envisioned that anything as severe as this could possibly occur. Yet it was now shown to be possible. Further, under certain circumstances, the diagonal winds could cause an increase in stress of more than 40 forty percent leading to the building beginning to vibrate severely.

LeMessurier looked at the design in detail and found that the weakest joint was on the thirtieth floor. If it failed, the entire building could collapse catastrophically. Weather analysis showed that New York historically encountered storms severe enough to cause a collapse as often as once every 16 years. A mechanical damper under the building improved the situation to a once-in-55-year probability, but the damper required electric current, something, which often failed in major storms.

LeMessurier had three possible courses of action—to remain silent, to report the problem and place the blame on the change in design made by the contractor, or to accept personal responsibility for the error. If he did remain silent, the probability was that sometime within the next 16 years, the building would fail and lives would be lost. However, the failure might never happen. Then again, if the building did fail, he didn't cause the problem. He designed the building with welded joints—not bolted ones. Someone else made the error.

He was facing a potential disaster. He chose to "blow the whistle" on himself and potentially face ruin—litigation, bankruptcy, and disgrace.

There was only one acceptable technical solution—go back and weld all of the bolted joints. But how? And hurricane season was approaching. LeMessurier followed the first canon of the NSPE *Code of Ethics* which states that the engineer shall "Hold paramount the safety, health and welfare of the public." In the words of the Creed he "placed service before profit, the honor and standing of the profession before personal advantage, and the public welfare before all other considerations." He reported his findings.

How was the problem corrected in a completed and occupied structure? Each night during the month of August 1978, a crew of drywall workers and carpenters entered the building at 5 PM, tore down sheetrock, and exposed some of the bolted joints. Welders started welding at 8 PM and worked until 4 AM. The carpenters, drywall workers, and laborers then cleaned up the mess so that the offices could be used normally during the day. Then on September 1, the worst fears of everyone came to pass---a major hurricane was headed toward New York. The work continued and enough repairs had been completed that it was believed that the building could now withstand a 200-year storm. Fortunately, that estimate wasn't tested because the storm turned away from the city. The repairs were completed and the building now can withstand a 700^{+}-year storm. It became one of the safest structures ever built.

What did it cost? According to LeMessurier, about $4.3 million. One construction firm claimed that they spent eight million and Citicorp declined to give their estimate. LeMessurier and his partners settled for two million dollars.

LeMessurier taught classes at Harvard University and talked about the experience with his students. He told them, "You have a social obligation. In return for getting a license and being regarded with respect, you're supposed to be self-sacrificing and look beyond the interests of yourself and your client to society as a whole. And the most wonderful part of my story is that when I did it, nothing bad happened."

Could you have acted as LeMessurier did?

STUDY QUESTION

6.1 In the Citicorp Tower case described above, LeMessurier's design was not faithfully followed. The joints were bolted without his knowledge. Under similar circumstances, if you had been the designer of the building would you have accepted the responsibility as LeMessurier did? It would be easy, and probably defensible, to claim that the error was the fault of the contractor, not the designer. Why in your judgement did LeMessurier accept responsibility? Instead of accepting responsibility, should he have taken action to require the contractor to accept responsibility and to perform the required repairs? List the pros and cons of both courses of action.

REFERENCE

1. J Morgenstern. The Fifty-Nine-Story Crisis. New Yorker, May 29, 1995: 45-53

7

What Constitutes Professionalism?

THE REAL WORLD OF ENGINEERING

I'd like to give you a picture of the real world as I encountered it when I graduated in chemical engineering from Carnegie Institute of Technology almost 40 years ago. Fresh out of college, I was very idealistic and ready to charge into the real world and set it right.

My ideals got set back very harshly and very quickly. First, I assumed that as a professional I could ignore the plant working schedule and come and go as I pleased as long as I got the job done. One day, I was caught up on my work and had little to do, so I went home early. That evening I got a call from the chief engineer informing me that under no circumstances did a professional employee do that. Instead, we were to be in the plant well before the union employees and were not to leave until after they were all gone—after all, we had to set the example. **Lesson #1**.

Then, after a few weeks of working in the plant, I came to the conclusion that it was really stupid to wear a coat and tie to work every day. After all, we never dealt with the public, only with construction workers and mill hands. Why then dress in good clothes in the grimy atmosphere of a steel mill? So I went to work one day wearing slacks and a sports shirt and promptly got chewed out by the chief engineer for not dressing like a professional. **Lesson #2**.

Not long after joining the company, we began working extended overtime and my pay checks began to skyrocket, whereupon I was promoted to the management ranks, given a modest increase in my base salary, told that I was now exempt from overtime pay, and that as a professional, I should be willing to work as many hours as necessary to get the job done and not expect extra compensation for it. **Lesson #3**.

Several years later, I left that company and took a job at West Virginia University, taking with me my industrial ideas of long working days, dressing as a professional should dress, and willingly working as long as necessary to get the job done. I encountered professionals of a different sort—professors—people who in the eyes of students may appear to work only a 3- or 4-hour day, dress in jeans and sports shirts, and wouldn't be caught dead in the office after hours or on a weekend.

Are the industrial people more professional than the professors? I think not. Professionalism isn't a visual appearance or a facade one puts on to impress others, nor is it a masochistic desire to sell your soul to the company store. Instead, it is an attitude—a desire to do a good job, to do it in an ethical and cost-effective manner, to strive to do an even better job in the future, to continue to educate yourself and expand your scope of knowledge, and to assist others who come along behind you to emulate what you do and know—and to do it better for the benefit of future generations.

There are many ways of developing a professional attitude and there is no set formula for doing so. Clearly, however, a necessary ingredient is participation and membership in the professional societies related to our individual fields of professional specialty. There really are three types of societies which professionals should participate in—the profession-wide societies, the societies in the various academic disciplines, and the specialty societies related to specific areas of work or career interests. The primary profession-wide engineering organization in the United States, of course, is the National Society of Professional Engineers and its constituent state societies. NSPE's concerns are those which affect all engineers regardless of discipline. NSPE is the political lobby of the engineering profession and works toward promoting effective legislation of benefit to the entire profession. NSPE also has many profession-wide programs of a nonpolitical nature.

The engineering disciplinary groups include the so-called engineering founder societies: the American Society of Civil Engineers; the American Institute of Mining, Metallurgical and Petroleum Engineers; the American Institute of Chemical Engineers; the American Society of Mechanical Engineers; and the Institute of Electrical and Electronic Engineers. The founders societies are the original group of disciplinary societies, those which joined together to establish the United Engineering Center and the Engineering Societies Library in New York.

These societies, and other disciplinary societies such as the Institute of Industrial Engineers, provide all manner of educational services through journals, textbooks, monographs, short courses, symposia, and perhaps most important of all, through their impact and influence on undergraduate curricula and accreditation of degree programs. Their efforts provide the broad underpinning of the various disciplines and the basic literature of each discipline. However, with the extreme diversification of the engineering profession, it is impossible for the disciplinary societies to provide the detailed information base required. We tend to become specialists and need specialized societies to serve our specific professional needs; to enable each individual engineering specialist to keep up-to-date on his or her field; and to provide the mechanisms for educating and training future generations of engineers. Thus, we have societies such as ASHRAE, the American Society of Heating, Refrigerating and Air Conditioning Engineers; ASQC, the American Society for Quality Control; ANS, the American Nuclear Society; and many others including another of my former employers, AACE International (the Association for the Advancement of Cost Engineering).

In my case, I quickly found out that my disciplinary society, the American Institute of Chemical Engineers, simply could not provide me with the support I needed in my narrow engineering specialty, cost engineering. So I sought out and became an active participating member of the specialty society which met my needs.

Does your society meet your technical needs? If not, seek out the specialty society which does. There are more than 100 listed in *Who's Who in Engineering*, one for every specialty imaginable. Seek out the one that fits your technical needs and become involved. It is your professional responsibility.

POLITICAL LOBBYING

As mentioned above, the National Society of Professional Engineers is the political lobby of the engineering profession in the United States and political activity at the federal level is a major part of NSPE's efforts on behalf of the engineering profession. The various NSPE-affiliated state and territorial societies serve a similar function at the state and territorial level and their chapters often engage in political activity at the local level.

At the January 1997 NSPE Leading Edge Conference in Charlotte, North Carolina, an interesting question was raised at one of the sessions by a conference attendee. In essence, this particular engineer questioned NSPE and state society activities in the legislative arena. He asserted that lobbying was unethical under the terms of the NSPE *Code of Ethics for Engineers*. Is it?

That question was presented to the Professional Engineers of North Carolina (PENC) Ethics Steering Committee which was meeting during the conference and was the subject of some discussion, as you would expect. After all, one of the primary activities of NSPE and its constituent state societies is lobbying. There was obvious concern therefore as to what ethical implications lobbying for the profession might have.

The discussion prompted me to review the *Code of Ethics* to see what if anything in it relates to lobbying activity. The conclusion I reached is that there is nothing in the *Code* which prohibits NSPE, the state and territorial societies, or any individual engineer from engaging in lobbying but there is much that affects the manner in which lobbying is done.

One of the "Fundamental Canons" contained in the *Code* is particularly relevant. It charges engineers to "... issue public statements only in an objective and truthful manner" and to "... avoid deceptive acts." Lobbying is by definition an effort to influence legislation and the outcome of legislation. This Canon makes it quite explicit that engineers are required to maintain objectivity and to be truthful in what they do. In the realm of politics, with public disenchantment of politicians running at an all-time high, engineers must be certain that they do not fall to the lowest common denominator when dealing in the political arena. We all know if we are not being truthful, but what about objectivity? It can be difficult to maintain objectivity when dealing with political issues which often become emotional.

The *Code* also states that "engineers shall hold paramount the safety, health and welfare of the public." Clearly this Canon is of primary import on any lobbying position which the profession takes relative to laws, codes, regulations, and other matters which may affect the public.

Read through the NSPE *Code of Ethics* in the Appendix and ask yourself if it prohibits lobbying activity. As you read the *Code* you will quickly realize that almost every item in it has a bearing on the manner in which engineers engage in lobbying. You should also quickly realize that the *Code* actually promotes and encourages lobbying rather than prohibiting it. Consider that the section on professional obligations mandates that "engineers shall at all times strive to serve the public interest" and shall "seek opportunities to ... work for the advancement of the safety, health, and well-being of their community." Encouraging effective legislation to regulate the practice of the profession is one way of doing this. Were it not for the lobbying efforts of NSPE and the state societies, much of the beneficial legislation which limits the practice of engineering to those who have demonstrated their competence would not be on the books. Nor would many rules and regulations which beneficially affect how facilities are designed.

We have not reached the ideal of uniform licensing or registration for engineers nor have we yet been able to assure that all of the rules and regulations which affect our work are reasonable and adequate. Until we reach this nirvana, we must continue our lobbying effort for the benefit of both the public and the profession. The NSPE *Code of Ethics* fully supports that objective.

ANOTHER FAILED BRIDGE

Chapter 6 discussed the Order of the Engineer and its evolution out of the failure of the Quebec City Bridge over the St. Lawrence River. In that failure, the girders supporting the center section of the bridge buckled, the bridge collapsed, and 76 men were killed. After redesign, construction began anew and the bridge failed a second time when its center span fell during construction. That bridge was eventually completed and remains in use today. It also stands as a reminder to engineers, in the words of Professor Henry Petroski of Duke University that, "The essence of modern engineering is to use our knowledge and experience to anticipate as completely as possible how any new design might fail and then take steps to avoid it" (1). He went on to say, "The engineer can calculate when failure will occur only if he knows what kind of failure to expect."

Successive bridge designs over the years have evolved using economy of materials and more slender aesthetic designs until a failure has occurred because the engineers did not fully understand or overlooked potential failure modes not present in earlier bridges having considerable overdesign.

The cycle has been repeated on many occasions. Before the Quebec bridge was the famous Brooklyn Bridge built by John Roebling and opened to traffic in 1883. It remains in use today carrying far greater loads and traffic than ever could have been imagined when it was designed. Another predecessor bridge was the massive Firth of Forth Railway Bridge built in Scotland in 1890, which also remains in use. The massive structure of this bridge is shown in Figure 7-1. Buckling could never be a problem with that massive structure. Designs evolved, however, and in 1907 disaster struck in Quebec.

Figure 7-1. Firth of Forth Railway Bridge. (Photo by Kenneth K. Humphreys.)

Prof. Petroski asks the rhetorical question, "Do engineers take excessive risks with human lives when they design bold new ways of doing things?" He goes on to say, "In the absence of something going wrong, successful designs are modified and changed until something does go wrong... [The engineer] can calculate those failure risks only after he has imagined how the structure can fail...This uncertainty is as old as engineering itself."

After the 1907 collapse at Quebec City, bridge designs were changed, many bridges were built, and designs kept evolving and "pushing the envelope" until 1940 when the Tacoma Narrows Bridge at Puget Sound near Tacoma, Washington was built. The bridge was opened to traffic on July 1, 1940. At 5,939 feet, it was longer and more slender than earlier bridges. It was two lanes in width with the deck supported on shallow steel girders instead of the open truss-work which was common on earlier designed bridges. The engineers were concerned about making the bridge stiff against bending but they overlooked the controlling failure mechanism, torsion. That oversight led to disaster. The bridge collapsed 4 months and 7 days after it opened.

Puget Sound is subjected to frequent gale force winds. These winds caused the deck of the new bridge to oscillate and twist when wind velocity increased. The bridge was a suspended plate girder bridge which caught the wind, rather that permitting it to pass through. Local residents nicknamed the bridge "Galloping Gertie" and some thought it was great fun to drive across the bridge when the winds picked up and the bridge started to twist and shake. Aerodynamic stability had not been a major factor in earlier, stiffer bridges but here it became the governing mode for potential failure.

Then on November 7, 1940 while Professor F. N. Farquharson from Civil Engineering Department of the University of Washington was filming the movement of the bridge, winds picked up to a gale force of 42 miles per hour, a not uncommon wind velocity in the area. The bridge began to move violently. Figures 7-2 through 7-6, photographs taken by Professor Farquharson, show in sequence what happened next.

Figure 7-2. "Galloping Gertie" twisting in the wind. (Courtesy of Special Collections Division, University of Washington Libraries. Photo by F. N. Farquharson, Negative # 4.)

Figure 7-2, a photograph, taken just prior to failure, shows the twisting motion of the bridge as winds picked up. When the twisting motion was at its maximum, Professor Farquharson calculated that the bridge twisted so severely that one sidewalk was as much as 28 feet higher in elevation than the other.

Figure 7-3, a photograph taken just before 11 o'clock in the morning, catches the first evidence of failure, concrete from the roadway dropping into the water. Bulges can be clearly seen in the stiffening girder near the far tower.

A few minutes later, about 600 feet of the roadway broke loose from the center span, twisted over, and crashed upside down into Puget Sound as is shown in Figure 7-4. The twisted girders can be clearly seen, as can a 25-foot section of concrete pavement falling away. Note the automobile on the remaining section of the center span. The remainder of the center span fell shortly afterwards, taking the automobile with it.

Figure 7-3. Portions of the Roadway Falling into Puget Sound. (Courtesy of Special Collections Division, University of Washington Libraries. Photo by F. N. Farquharson, Negative # 2.)

Figure 7-5 shows the west span of the bridge after the failure. Note the sag in the span due to the loss of counterbalancing weight of the center span. Note also the twisted remains of the north stiffening girder dangling into the water.

The next photograph, Figure 7-6, shows the east span after the failure. The dangling remainder of the south stiffening girder and the broken end of the north girder are clearly shown.

When the Tacoma Narrows Bridge collapsed, several other bridges on the drawing board were redesigned with greater margins of safety. The Tacoma Narrows Bridge was rebuilt and opened to traffic ten years later on October 14, 1950. The bridge was rebuilt with the same type of structure as the original bridge but with a deep open truss deck. Wind tunnel tests were made on the design to verify that unusual wind speeds could not resonate the bridge.

The replacement bridge, shown in Figure 7-7, is 5,979 feet in length, 40 feet longer than the original bridge. The 2,800-foot center span stands 188 feet over the water.

Figure 7-4. Center span crashing into Puget Sound. (Courtesy of Special Collections Division, University of Washington Libraries. Photo by F. N. Farquharson, Negative # 12.)

Figure 7-5. West span of the Tacoma Narrows Bridge after the failure. (Courtesy of Special Collections Division, University of Washington Libraries. Photo by F. N. Farquharson, Negative # 7.)

Figure 7-6 East span of the Tacoma Narrows Bridge after the failure. (Courtesy of Special Collections Division, University of Washington Libraries. Photo by F. N. Farquharson, Negative # 9.)

The Mackinac Bridge spanning the five-mile-wide Straits of Mackinac between the Upper and Lower Peninsulas of Michigan was under construction in 1940 when the Tacoma Narrows Bridge collapsed. It had a similar design and work had begun on its approaches. Work on the Mackinac Bridge was halted while the bridge was redesigned. World War II intervened and it was not until 1957 that the Mackinac Bridge was completed (2). It is shown in Figure 7-8 and looks quite similar to the new bridge at Tacoma Narrows.

The Mackinac Bridge is considerably longer than the Tacoma Narrows Bridge at 8,614 feet including the anchorages. Its center span is 199 feet above the water and the total length with approaches is five miles. Wind forces at the Straits of Mackinac are often much more severe than those at Puget Sound. Imagine the consequences if construction had proceeded with the original design.

From an ethical point of view, it is unclear if the designers of these bridges did anything wrong. They used commonly accepted designs but they neglected the potential failure mode of these bridges, torsional stresses caused by high winds. Their error was one of omission, not one of commission. As Professor Petroski said, "Without engineers there would be no disasters. We depend upon engineers for new constructions, to provide us with safe shelter, safe transportation, and safe power. We tend to take their successes for granted but their failures are headline news" (1).

Figure 7-7. Replacement Tacoma Narrows Bridge. (Courtesy of Special Collections Division, University of Washington Libraries. Photo by F. N. Farquharson, Negative # 5.)

Figure 7-8. Mackinac Bridge as seen from Mackinaw City, Michigan. (Photo by Kenneth K. Humphreys.)

William J. LeMessurier did not at first "...imagine how the structure can fail" in the design of the Citicorp Tower in New York. Fortunately he did eventually discover the governing failure mode before a disaster occurred. The engineers on the Tacoma Narrows Bridge project were not as fortunate.

COMPUTER-AIDED NIGHTMARES

When I first began to use computers more years ago than I care to remember, off-the-shelf software was something you only dreamed about. There was no cadre of readily available programmers and there was no convenient computer supply store where you could buy a software package to do virtually anything. There was only yourself and your ability to write your own programs. For an engineer that usually meant mastering FORTRAN (FORmula TRANslation), a computer language created for engineers to convert our mathematics and formulas into something the computer could understand.

You painstakingly analyzed the problem at hand, decided the method of solution, sat down and wrote out the mathematics and formulas, and then converted that to the FORTRAN language. Next you punched it all into a stack of cards and read them into the computer which hopefully did what you wanted, but more often than not didn't. You usually had a long sequence of try, debug, try, debug, and try again before you finally had a workable program that did what you wanted. It was indeed a tedious proposition!

However, when you did finish writing and debugging the program, you knew what it contained. You knew what it assumed, what engineering principles applied, and when it was valid to use the program. You also knew when the program was not valid. It was an extension of yourself—it contained your engineering experience, judgment, and expertise.

Today, we are far beyond that. Engineers rarely write their own programs anymore. We have programming staffs to do that or, more likely, we buy a software package from one of the many very qualified suppliers in the business. A vast improvement isn't it? **Not necessarily so!**

One of the problems with software obtained from others is that, no matter what the documentation or salesperson says, you can't be sure that it will always apply to your situation. You don't necessarily know the inherent assumptions which went into the design of the software, nor do you know with certainty what its limitations are. These computer programs and the computers themselves are wonderful tools, **but don't use them blindly**. Question what the software produces. Don't take the output on faith. It might not be correct. The software might assume "average" conditions and your particular project might deviate from average (and probably does).

Your name goes on the project, not the name of the anonymous person who wrote the program. If the design is flawed, you are responsible—not the programmer. You have the professional and ethical responsibility to check the results. Examine them. Are they reasonable? Do they fit the project conditions? Do they make sense? Do they coincide with what your experience and judgement suggests? They should, and if they don't, find out why.

Not long ago, I was asked to review an impressive engineering report prepared for a client of mine by a major U.S. engineering and construction firm. I looked at it and something just didn't ring true. My experience waved a big red flag to me so I asked to meet with the engineers who had prepared the report. I learned that virtually everything in the report had been generated using a very widely known and respected piece of software, a software package that is generally accepted as being very reliable. *This time it wasn't.*

Why? The package assumed a certain set of design conditions for a project unless the input specified that different conditions applied. Unfortunately the documentation did not clearly explain the default assumptions built into the software. In this project, many things were different but the input had not been modified accordingly. The result was that the report was totally flawed—the program simply was not applicable to the project at hand. The E & C firm should have known this but they didn't. They were comfortable with that software and had grown complacent in using it. The engineers assumed it was valid. They didn't check the results for reasonableness and they got burned.

That E & C contractor lost the job, one that was worth over $500 million. I don't know but I would not be surprised if some engineers also lost their jobs—**they should have.** In this case, blind faith in a piece of software cost the firm a lot of money. In another set of circumstances, this type of error could have cost lives.

A COMPUTER-ASSISTED CATASTROPHE

Prior to the advent of the digital computer, engineers used slide rules for design calculations. The slide rule was the symbol of the profession and engineering students at any university could usually be visually identified. They wore a leather holster on their belts containing a slide rule. The most common slide rules were about ten inches in length and no engineer would ever be without one. They were vital tools of the profession.

My slide rule was a Keuffel & Esser "Log Log Duplex Decitrig". It was a beautiful instrument made of what appeared to be mahogany with white enamel or ivory faces. As shown in Figure 7-9, it had nine scales on one side and eleven on the other. With it I could perform multiplication and division; calculate squares, square roots, cubes and cube roots; do trigonometric calculations; and do logarithmic calculations using logs to the base e or base 10. It was a marvel and it never needed batteries or an electric outlet. It was totally reliable and never gave me the message that, "This program has performed an illegal operation." All that was required to keep it from "locking up" or "freezing" was an occasional application of some wax on the sliding center section.

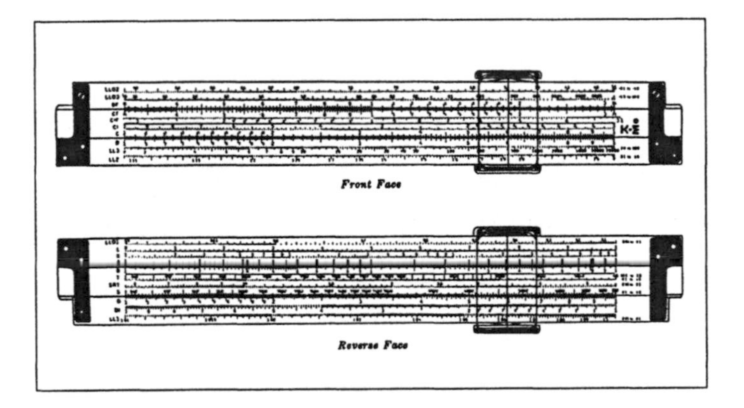

Figure 7-9. The Keuffel & Esser Log Log Duplex Decitrig slide rule.

Unfortunately, the slide rule was rendered obsolete, first by the scientific calculator and then by the computer. The speed and precision of these electronic marvels quickly displaced the slide rule, which at best could only give an answer to about three significant figures with any degree of accuracy at all. That wasn't a problem however because it was, and still is, sufficiently accurate for most engineering applications.

The slide rule did not tell you where the decimal point belonged. You had to figure that out for yourself. You had to know the expected magnitude of your answers. I never found that to be a problem. If you knew what you were doing, you had a good sense of the magnitude of the correct answer. Today, computer answers are all to often accepted at face value. They often aren't questioned, as is illustrated by the example of the E & C contractor discussed above.

Slide rule calculations took a lot more time than computer calculations do today. Complex designs were difficult and time consuming to perform. Consequently, many projects which are routine today would never have been attempted in the slide rule age. Computers are so fast that they can perform calculations in a fraction of a second, which would have taken months or years with slide rules.

However, the computer cannot think. It is nothing more than a very fast, highly accurate machine that can do no more than human beings program into it. The computer will simply spit out an answer that would have been questioned by the slide rule engineer based upon his or her experience. The computer won't look at the answer quizzically and say, "Is this reasonable? Have I thought of everything?" Unfortunately, in the words of Duke University Professor Henry Petroski, the computer "...gives us more information than we are able to assimilate. It gives the illusion of looking at everything. It also gives more apparent precision than can be meaningful...I believe that we should never rely entirely on the computer to anticipate failure. In particular, we should not expect it to think of failure modes that human engineers and programmers haven't already thought of." (1).

Professor Petroski goes on to say that, "Human engineers worry and lose sleep..." thinking about the consequences of their designs. Computers don't worry. They just do what has been programmed into them—even if the programming has overlooked something, is incomplete, or is inadequate. Computers are fast, accurate, and can generate a false sense of security.

One classic example is the collapse of the computer-designed roof of the Hartford, Connecticut, Civic Center in 1978. This was a complex roof structure that probably would never have been attempted before the advent of computers. It was about 2-1/2 acres in area and was supported by only four columns. It provided spectators with an unobstructed view of the arena floor.

In January 1979, thousands of people attended a basketball game in that arena. It was a cold night and snow was falling. The roof also fell—fortunately after the game was over and the spectators had left. It was unable to support the weight of the accumulating snow and ice. If the snow had been coming down faster, the roof might have collapsed during the game killing and injuring thousands of people.

The roof collapsed because the rods supporting it had buckled. This mode of failure was completely unanticipated by the computer. Professor Petroski commented, "The design would have been impossibly tedious for the slide rule engineer." The project probably would never have been undertaken except for the promised speed and reliability of the computer. Professor Petroski referred to this type of computer design problem as "...a new mode of failure, the computer aided catastrophe."

The point of this discussion is that engineers, not computers, are responsible for the consequences of their designs. It is not sufficient to design by computer and proceed to build. The engineer has the ethical obligation to do everything possible to verify the suitability of the design for its intended purpose. The first Fundamental Canon of the NSPE *Code of Ethics* must remain ever present in the engineer's mind. The engineer should never blindly accept a computer-generated design without question.

The title of this chapter is "What Constitutes Professionalism?" It discusses several things that are a partial answer to that question. The most important part of the answer is the first Fundamental Canon which says, "*Engineers, in the fulfillment of their professional duties, shall hold paramount the safety, health, and welfare of the public.*" Computers can't do that. It is a personal responsibility.

STUDY PROBLEMS

7.1 What is the disciplinary society that serves your field of engineering study? What are its requirements for membership? What member services does it provide and what are its publications? Contrast its services and publications with those of the National Society of Professional Engineers.

7.2 If you are working in a particular engineering specialty, or if you plan to work in a specialty field, what is the disciplinary society which serves that specialty field? What are its requirements for membership? What member services does it provide and what are its publications? Contrast its services and publications with those of your disciplinary society.

7.3 What are the three major types of engineering societies and why should engineers seek membership in all three?

7.4 Do you agree or disagree with the author's conclusion that political lobbying is both acceptable and beneficial to the engineering profession? Why or why not?

7.5 In your opinion, did the designers of the Tacoma Narrows Bridge act ethically? Did they use an acceptable standard of care in the design? If not, what could they have done to verify that their design was safe?

7.6 What type of actions should an engineer take to assure that computer-aided designs are reasonable and have included proper consideration of the potential failure modes?

7.7 Consider the following case from the "You Be the Judge Column" of *Engineering Times*:

Mann Edgingrisk, PE, performs hazardous waste remediation services for his clients. The contract he uses with the clients has an indemnification clause that requires the client to "indemnify and hold harmless Edgingrisk for any damages or legal costs (including attorneys fees) arising from the performance of the hazardous waste remediation services." However the contract also provides that the indemnification would not apply to losses resulting from Edgingrisk's "willful misconduct or sole negligence." Edgingrisk included the indemnification clause after carefully studying the hazardous waste insurance market and determining that the currently available insurance coverage does not adequately protect him and his firm.

Was it ethical for Edgingrisk's contracts to require clients to indemnify and hold him harmless for any damages or legal costs (including attorneys fees) arising from the performance of the hazardous waste remediation services?

REFERENCES

1. H Petroski. To Engineer is Human. Videotape from the BBC television series Horizon. Distributed in the United States by Films Illustrated, a subsidiary of Public Media Films Inc.; distributed in Canada by BBC Television Distributors.
2. The Mackinac Bridge. Information brochure, Mackinaw Bridge Museum, Mackinaw City, MI, 1998.

8

Working Within Your Discipline*

One of hardest decisions an engineer sometimes has to make is the decision to turn down paying work, whether you work for large corporation or are self-employed. However, the NSPE *Code of Ethics* clearly demands that we say no when we are asked to work outside our area of competence. This is a troublesome area for two reasons. First, most of us have a hard time saying no. Second, the bounds of our expertise are rarely well defined. Most of us have at least some experience, albeit limited, outside our primary practice area. So it is left up to us to determine the limits of our competence. The trick is knowing where to draw the line.

ENGINEERS WEAR MANY HATS

Many engineers work for a relatively small company, in which they may be one of only a few, or possibly the only, engineers in the company. In such a situation, an engineer is required to wear many hats, and most of the time this does not cause trouble. In fact, our engineering education was probably designed to train us for these situations. That's why engineering schools require mechanical engineers to take courses in electrical engineering, and vice versa. The reasons are obvious. In many cases our job *requires* us to practice in these gray areas.

As professional engineers, however, we have to be careful about overestimating our abilities. While an employer can reasonably expect an engineer trained in mechanical engineering to handle incidental electrical circuit design, he certainly should not be expected to design complex electrical or electronic components, or to handle a large civil/structural design project. Such work should be assigned to an engineer trained and experienced in the specifics of electrical or civil/structural engineering. Many employers may not understand the distinction, so part of our job as an engineer may be to educate others about the discipline-specific nature of engineering. The same applies to clients in private practice situ-

---

* By Russell C. Lindsay, PE, PARC Engineering Associates, Asheville, North Carolina

ations. Certainly, graduate engineers specializing in fields like software and computer engineering are probably poorly qualified to design a bridge, or even the electrical system for a building.

ACADEMIC KNOWLEDGE AND EXPERIENCE

One area to watch out for when working in the "gray area" between disciplines is our lack of experience with the codes and standards applicable to areas outside our primary practice area. This includes safety codes for products, material specifications, etc.—not just building codes. Unless one works with these codes and standards regularly, a significant amount of study and research may be required to properly understand their application in a particular situation. Assuming that time spent to do this research will be billed to our employer, or to a private practice client, it could easily be cheaper to hire a specialist in that field from the outset. Also, the "language" of engineers is very different from one discipline to the next. A welding symbol, for example, might be misinterpreted by an electrical engineer, resulting in a structural collapse due to an improper weld. Thus, while we *may* be competent to expand our practice into another discipline (with sufficient study), we may also do our employer and/or client a financial disservice by agreeing to practice in this manner. Worse, our lack of real-world experience in the "other" discipline may endanger the public safety, or our company's future.

Note that many engineers change their technical emphasis and/or expand their area of expertise over the course of their career. The recent transition of large numbers of engineers from the military-industrial complex to more commercial enterprises is one case in point. Some do it gradually, by working on a team with someone experienced in another field and studying on their own or in seminars and short courses. Others may move more rapidly—in a structured degree program or by changing jobs and acquiring intensive on-the-job experience. Engineers may also become licensed surveyors, architects, lawyers, or business managers during their career. Certainly we should remember that engineering licensing boards consistently require extensive experience before granting engineering licensure to a new engineer. Comparable experience should therefore be expected in any new discipline in which an engineer chooses to offer professional services. Some states even track discipline-specific licensure.

Some engineers, by virtue of their wide-ranging duties in an industrial or construction environment, become experienced in many disciplines, but may have only a limited depth of knowledge in any of them. Obviously, these "generalist" engineers are making a valuable contribution to their company and to society. However, they must be careful not to overestimate their abilities and should call on specialist engineers and other professionals as warranted. A generous dose of professional judgement is clearly required. Anything else may endanger the welfare of the public or the future of their employer.

AN ENGINEER IS AN ENGINEER, RIGHT?

Another pitfall is the friend or business associate with a small project requiring engineering assistance. Many times the project would be completed without engineering input if not for local codes requiring an engineering seal on the plans. Often the "plan" is presented to an engineer as a completed document, prepared by a draftsman or an unlicensed engineer, needing only an engineering seal to be accepted for construction. The request seems simple enough. "You are an engineer, so could you review this plan and approve it so that I can get it built? I've checked with the big engineering firm downtown, and they want $3000 to draw up a set of plans for me. I already have the plans drawn. All I need is an engineering seal. I know you are an electrical engineer, but you took courses in structural design, right?"

One problem with this situation is that even if the project was in the engineer's discipline, the plan was not prepared under his supervision and control, often called "responsible charge." Approving such a plan is a direct violation of the engineering licensing statutes in most states, and is often called "plan stamping." Yes, the alternative appears more time-consuming and costly. However, the idea is that engineering plans should be just that—plans executed by an engineer—not plans that have been "reviewed" by an engineer. Think about the thought processes involved in executing a plan from scratch versus that of reviewing someone else's plan. The depth and breadth of the thought process is much different. Also, placing your signed "seal" on someone else's plans is, quite possibly, a form of plagiarism. Even if it is done with the original designer's permission and with the designer clearly identified on the plans, this practice is misleading and potentially confusing to the public officials and contractors using the plans. Who should they call for clarification regarding the plans? And who approves changes and is liable for errors? Protection of the public, the basis for requiring the engineering "seal" in the first place, is only afforded when an engineer qualified in the relevant discipline is responsible for all aspects of planning, design, and field supervision of the project.

Example

Suppose electrical engineer Dee C. Current was asked to review and "seal" the structural plans for a deck to be added to her neighbor's house. The proposed design includes tall support posts and a cantilevered section due to a steeply sloping lot. She might feel comfortable performing a few structural calculations to check the beam strength, checking the building code for foundation requirements, and reviewing the hardware specifications. She probably would not examine other configurations or be well versed on column buckling or seismic design principles, or special loadings such as snow or wind. And she is not likely to question the basic configuration. A civil/structural engineer, on the other hand, would begin the design process by examining the load requirements, including seismic, wind, and snow loads, and his experience might suggest a more efficient, less costly, or even safer design based on knowledge of local soil and environmental conditions and other factors. He would be knowledgeable about the hardware specifications needed to resist corrosion and other failure modes in that local area. The design might look the same to the electrical engineer, Dee. However, inconspicuous details could be the difference between a safe, long-lasting structure and one that collapses under the next heavy snow, high tide, windstorm, or neighborhood party.

Actual Case Study

A driver loses control of her car while driving on a straight and level road, resulting in a rollover and serious injuries to the driver. She reports a sudden and unexpected loss of control, for no apparent reason. After the collision, one of the rear wheels is found separated from the car, the stub axle broken off. This leads the driver to conclude that either the tire blew out or the axle broke, causing the loss of control. The automobile insurance company requests an initial investigation by a mechanical engineer experienced in automobile collision reconstruction and mechanical failure analysis. This engineer finds evidence of impact forces applied laterally to the side of the wheel rim, obviously during the rollover. The wheel-bearing race also contains distinctive imprints made by the bearings, consistent with a severe side impact. A close inspection of the fracture surface reveals no evidence of fatigue failure, the most likely mode of axle failure while driving. Clearly, he concludes, the axle broke *during* the rollover, not before.

Months after the collision, the car driver and her attorney hire a *mining* engineer to inspect her car and reconstruct the incident, in hopes of filing a product liability lawsuit against the car manufacturer. He examines the car, sees the broken axle, now rusted, and agrees that her conclusion is logical. [Car wrecked + axle broken = broken axle caused wreck.] Luckily, a telephone conversation with the first engineer alerts him to the telltale signs he overlooked. If not for this phone call, the "out of his field" mining engineer might have encouraged the car driver and her attorney to spend large sums of money pursuing a frivolous lawsuit destined to failure.

Was it ethical for the mining engineer to accept such an assignment?

CONCLUSION

Engineering is an exciting field, and one of the best things about it is the wide range of career opportunities it affords. However, as professionals, engineers must remain vigilant so that they do not overestimate their abilities in fields outside their chosen discipline. Expanding their area of practice is fine, as long as they keep the first two "prime directives," Fundamental Canons of the NSPE *Code of Ethics*, in mind. These two Canons say that, "Engineers, in the fulfillment of their professional duties, shall:

1. Hold paramount the safety, health and welfare of the public.
2. Perform services only in areas of their competence."

STUDY QUESTION

8.1 The "You Be the Judge Column" of *Engineering Times* presented the following case:

E. E. Most, a PE in the electrical engineering field, is employed by a state agency as a computer systems engineer with some management responsibilities. Educated and trained to perform customary engineering services, Most's work experiences have never involved technical and design issues concerning environmental services. As part of a restructuring of the agency, his direct supervisor, Felix A. Ball, recommends that Most accept a position the Department of Environmental Services has offered him. The position requires a PE and involves engineering analysis and design responsibilities, and Most would be working as part of a team of engineers.

Most refuses to accept the position, citing state board regulations requiring him to perform work only in his area of competence and his lack of experience to perform the work. Thereafter Most is terminated.

During an administrative hearing involving Most's reinstatement to his former position and back pay, Ball testifies that Most was qualified to accept the position offered in the Department of Environmental Services.

What do you think? Was it ethical for Most to decline the position? And did Ball act ethically in testifying that Most was qualified to accept the position?

9

Reasonable Safety vs. Foolproof Design*

It seems that almost every week we hear about a safety recall of a consumer product. The frequency of these reports and the sensational nature of the news media feed the perception that design flaws occur quite frequently in our high technology world. Maybe they do, but they certainly shouldn't, especially if the designers are competent, professional engineers.

The NSPE *Code of Ethics for Engineers* and the engineering statutes in most states clearly delineate an engineer's duty to "hold paramount the safety, health and welfare of the public." This can be construed to include many aspects of economic as well as physical welfare. However, this chapter is devoted to safety, defined in my dictionary (1) as "the condition of being safe from undergoing or causing hurt, injury, or loss." And the word *safe* is defined as "freed from harm or risk," implying an absolute condition much like the case of a runner called "safe" in the game of baseball. In the real world, however, there is no such thing as absolute safety, i.e., without any degree of risk. Absolute safety is a little like the mathematical concept of infinity—we recognize that it exists, but we know we can never reach it. As engineers, therefore, we are really concerned with the concept of "reasonable safety." Extending the mathematical analogy, reasonable safety is like an approximation of a mathematical function using the first few terms of an infinite series. The more terms we use, the better the approximation. But how many is good enough?

A number of good reference texts are dedicated to the subject of safe design. Some of them (2,3,4) indicate that "reasonable safety" is defined largely in terms of our legal system and the litigation environment. If a product causes, or is a proximate cause of, an injury or death, there will almost certainly be some type of financial settlement between the parties involved in the design and construction of

----

* By Russell C. Lindsay, PE, PARC Engineering Associates, Asheville, North Carolina

the product and the injured person(s). Product in this sense could mean an engineered structure or facility, as well as a consumer or industrial product. If a settlement cannot be reached outside the legal system, a lawsuit will usually result. This will put the issues and judgements in the public record and make them available as "precedent" for future cases of a similar nature. Thus our duty to design reasonably safe products and structures is dictated both by our *responsibility to the public* and our *responsibility to protect our employer (or client) from liability exposure and/or the fallout from negative publicity*.

So why is this an ethical issue? And how can engineers make sure that the products, structures, and designs they produce will be reasonably safe in the modern world?

The concept of reasonably safe design is an ethical dilemma because of the potential for conflict between a professional engineer's responsibilities to the public and to an employer or client. How can we design safe products without bankrupting our company? How much is it worth, in time, money, or other resources, to save a single human life? What about the environmental issues of air, water, etc? In each of these cases, we must find a balance, a "reasonable" middle ground, between two sometimes-distant extremes. But we don't have to reinvent the wheel. If we look closely, there are road signs on this sometimes-twisting path.

LEGAL STANDARDS FOR LIABILITY

In order to act as "faithful agents or trustees" of our employers and clients, engineers must be familiar with the legal liability landscape. Product liability law rests primarily on three legal standards or theories—negligence, strict liability, and express warranty. Briefly put, *negligence* is a standard that tests the *conduct* of a defendant, such as an engineer or a company that designed a product. *Strict liability* tests the *quality of the product*, balancing risk versus utility in the real world. *Express warranty* tests whether a product *performs* in the manner represented by the manufacturer, regardless of whether there is a proven defect in the product.

In negligence cases, courts have generally ruled that if a product is inherently dangerous and found to be defective, a manufacturer is liable for injuries to a user provided (3):

1. The defect caused the injury;
2. Reasonable means existed to prevent the defect; and
3. The user used the article in a reasonable manner.

As you can see, the term "reasonable" is used to refer both to the designer's choices and actions and to the actions of the user. This intentionally vague wording allows for flexibility in its application to the wide variety of products in the marketplace and the diversity of cases presented to the court system. Furthermore, the court will usually be asked to consider three factors: the *probability* that harm will occur; the *gravity* of the harm; and the *burden of taking precautions* to protect against the harm. A "standard of care" approach is generally used in evaluating these factors with respect to a negligence claim against a professional. This standard is often defined as "that degree of care, which a reasonably prudent person [engineer] should exercise in same or similar circumstances." It clearly does not expect perfection, and errors of judgement are not considered negligence in most cases.

In a strict liability action, a manufacturer can be held liable if it places a defective product, one that is judged *unreasonably dangerous*, in the marketplace. The focus in strict liability cases is on the product, not the conduct of the company or designer. However, the broad nature of this legal theory requires engineers to be especially diligent in their design efforts so as to protect the interests of their employers or clients. Table 9-1 lists some measures proposed for evaluating the risk versus utility of a product to determine whether that product is unreasonably dangerous. Again, perfection is not necessarily required, but a company can be held liable for a defective product even if the defect or failure was completely unforeseen, or impossible to predict.

Finally, express warranty actions are based on the warranties or representations that accompany a product or precede it in advertising. Courts have ruled, for example, that terms such as "gentle," "harmless," or "blow-out proof" constitute express warranties. A related concept is the implied warranty of merchantability, i.e., the act of selling a product implies its suitability for the marketplace. Implied warranties originated in the common law, but the concept is now codified in the Uniform Commercial Code. These types of liability require close coordination between engineering and marketing departments to prevent unrealistic product claims or substandard performance that could result in a product liability "warranty" action.

Engineers should not necessarily define their practice by the limits of liability case law. However, ethical practice requires us to understand the basic principles if we are to protect the interests of our clients and employers. Also remember that the legal liability standards have evolved to protect the public—the same public that the NSPE *Code of Ethics* demands that we protect. The following pages summarize some of the procedures that can help us in that effort to protect both the public and our employers/clients. While most of this discussion centers on "product" design, similar procedures apply to civil, structural, environmental, and other types of engineering design.

IDENTIFICATION OF HAZARDS AND "DESIGNING THEM OUT"

The first step in executing a reasonably safe design is recognizing that we have an ethical duty to do so, and then looking for the hazards. The safest design is one in which the designer keeps safety in mind from the start. And despite the nature of most engineering education, which typically stresses the hard science aspects of design and analysis, consideration of human factors is essential in design of reasonably safe products. Consider using a systematic approach, possibly using the "guidelines for designing goofproof products" in Table 9-2.

Table 9-1. Indexes for Determining if a Product Is Unreasonably Dangerous

How useful and desirable is the product?

Are other, safer products available that will meet the same needs?

Is the product likely to cause injury, and is the injury likely to be severe?

How obvious is the danger?

How much knowledge about the danger does the public have, and what could it normally expect regarding the danger (particularly with respect to established products)?

Would care in using the product (i.e., following instructions or heeding warnings) avoid the injury?

Could the danger be eliminated without seriously impairing the usefulness of the product or making it unduly expensive?

Source: Reproduced by permission of ISI Publications, Inc. from S Brown, ed., Forensic Engineering: Part I, An Introduction to the Investigation, Analysis, Reconstruction, Causality, Risk, Consequences, and Legal Aspects of the Failure of Engineered Products, 1995, ISBN 0-964-553600 (adapted from W Dean. Strict Liability of Manufacturers. Southwestern Law J 19(5), 1965).

Table 9-2. Principles for Designing Goofproof Products

1. Recognize and identify actual or potential hazards, then design them out of the product.
2. Thoroughly test and evaluate prototypes of the product to reveal any hazard missed in the preliminary design stages.
3. Make certain the product will actually perform its intended function in an acceptable manner so that the user will not be tempted to modify it or need to improvise possibly unsafe methods for using it.
4. If field experience reveals a safety problem, determine its real cause; develop a corrective action to eliminate the hazard, and follow-up to make certain that the corrective action is successful.
5. Design equipment so that it is easier to use safely than unsafely.
6. Realize that most product safety problems arise from improper product use rather than product defects.

Source: Reproduced by permission of The McGraw-Hill Companies from TA Hunter. Engineering Design for Safety. McGraw-Hill, 1992, p. 77.

Remember that the public is not always aware of the hazards to which they are subjected. Engineered products frequently store energy—sometimes kinetic, electrical, chemical, or potential energy. The general public often does not appreciate the hazards of stored energy; especially those associated with rotating machinery, electrical components, and many other types of engineered devices. As engineers, we are better equipped to recognize these hazards in our designs and to eliminate or minimize them. And it is considerably easier to recognize and consider the hazards early in the design process than to redesign the product after it is in production or under construction.

There are a number of rigorous methods for analyzing the risks or hazards of a design. Table 9-3 lists some that may be useful during either the initial design of a product or during an analysis of a product failure or injury accident after it occurs. Many other types of analysis, such as *An Instructional Aid for occupational Safety and Health in Mechanical Engineering* (5), have been developed or adapted for specific industries or disciplines.

Once all of the hazards in a particular design have been identified, the logical next step is to rank them. Evaluating and prioritizing hazards is necessary because we have an ethical responsibility to be good stewards of our employer's, our client's, and the public's resources. Ranking of hazards may be somewhat subjective, especially if little or no statistical information is available. However, even a subjective ranking is better than no ranking. Hazards are usually ranked according to two primary factors: the *frequency of exposure* and the *severity of the risk*. By combining the frequency and severity rankings one can obtain an overall risk factor for each of the identified hazards. Note that a hazard with little or no chance of causing a fatal injury can be ranked higher than a hazard capable of resulting in fatal injury—*if* it has a much higher frequency of exposure. This ranking, imperfect as it may be, at least provides a guideline for prioritizing our hazard elimination efforts. It makes little sense to expend valuable resources on a hazard near the bottom of this ranking until the hazards above it have been corrected.

Another step that many engineers forget is identifying foreseeable misuses, and thus additional foreseeable hazards, of their product. Not everyone will use a product in the manner intended, and if the design makes it easy to use unsafely, you can bet that more that one person will use it that way. Some people call this Murphy's Law, illustrated in Figure 9-1 as it applies to assembly of a mechanical device. Because users may not fully comprehend the hazards, engineers must take responsibility for identifying foreseeable misuses and the hazards that result. A design feature that appears safe may become a serious hazard under some types of reasonably foreseeable misuse. Again, consider using a systematic method for discovering and evaluating reasonably foreseeable hazards, possibly using a checklist approach, a brainstorming approach with a systematic follow-up evaluation, or some similar technique (4).

Table 9-3. Types of Analysis for Evaluation of System Hazards

Gross-hazards analysis, which is performed early in a design stage and considers overall system performance instead of individual components.

Classification of hazards identifies the types of hazards disclosed in gross-hazards analysis and displays them according to potential severity.

Failure-mode and failure-mechanism (mechanics) analysis.

Fault-tree analysis, which outlines the possible sequences of events leading up to an incident.

Energy-transfer or energy-release analysis, which determines the interchange of energy that occurs during a catastrophic event.

Catastrophic analysis, which identifies the modes of failure that would create a catastrophic accident.

Systems-subsystems analysis, which reveals the interfaces between systems.

Maintenance hazards analysis, which evaluates the performance of a system from a maintenance standpoint and addresses whether the maintenance procedures produced new hazards.

Human-error analysis, which defines the skills required to operate and maintain a system, considers the question of whether a failure could be initiated by human error, and if so, how the error affects the system.

Transportation-hazards analysis determines the hazards to shippers or innocent bystanders.

Consequence analysis assesses the consequences of a system failure with respect to personnel injury, property damage, facilities loss, litigation (real cost), and public relations (indirect losses).

Hazardous-release-protection analysis considers the design of protective systems to reduce energetic (blast, missiles, ground motion) and degenerative (radioactive, chemical, biological, horrible) hazards to acceptable levels.

Source: Reproduced by permission of ISI Publications, Inc. from S Brown, ed., Forensic Engineering: Part I, An Introduction to the Investigation, Analysis, Reconstruction, Causality, Risk, Consequences, and Legal Aspects of the Failure of Engineered Products, 1995, ISBN 0-964-553600.

With the hazards identified and ranked, engineers can then take or recommend action regarding design choices with respect to these hazards. Obviously, the best course of action is to design the product so as to eliminate the hazard. We do this when we design a building for all foreseeable combinations of environmental conditions and occupant loads, eliminating the risk of structural collapse. Sometimes, the hazard is inherent in the function of the product and cannot be entirely eliminated, such as the pinch point or point-of-operation hazards in presses and other production machinery. In these cases, guards or other features designed to physically separate users from the hazard are the next best approach. The Occupational Safety and Health Administration (OSHA) literature contains numerous publications detailing the principles of machine guarding. These publications can often be obtained, sometimes free of charge, from your state's Department of Labor.

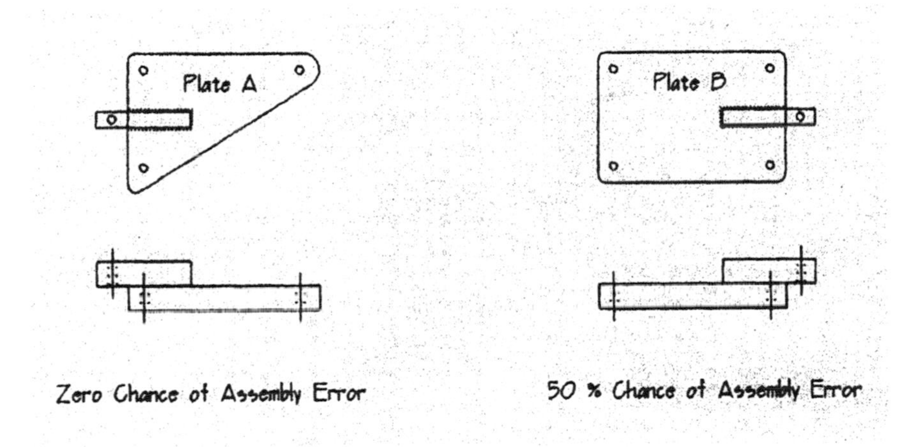

Figure 9-1. An Example of Murphy's Law. (Source: Adapted from Ref. 5.)

With the hazards identified and ranked, engineers can then take or recommend action regarding design choices with respect to these hazards. Obviously, the best course of action is to design the product so as to eliminate the hazard. We do this when we design a building for all foreseeable combinations of environmental conditions and occupant loads, eliminating the risk of structural collapse. Sometimes, the hazard is inherent in the function of the product and cannot be entirely eliminated, such as the pinch point or point-of-operation hazards in presses and other production machinery. In these cases, guards or other features designed to physically separate users from the hazard are the next best approach. The Occupational Safety and Health Administration (OSHA) literature contains numerous publications detailing the principles of machine guarding. These publications can often be obtained, sometimes free of charge, from your state's Department of Labor.

The design features used to eliminate or guard against hazards are often very industry and discipline specific, so they will not be discussed here. However, the universal aspect of this process is one of deciding which design features or modifications will be included in the final design. This is a complex decision-making process, involving not only design engineers, but also management, manufacturing, and financial concerns. Often, engineers do not have ultimate decision-making authority. Yet, engineers have a responsibility to their client or employer to "act as faithful agents or trustees." The explanation for this responsibility in the

Code of Ethics is primarily directed toward financial conflict of interest situations between clients/employers and other parties. However, ethical concerns may also arise when the economic or other interests of the client/employer are in conflict with public safety concerns. The engineer's goal should be to reach a reasonable balance, holding safety paramount and still protecting the economic interests of the client or employer. If such a balance cannot be reached, the first fundamental canon, i.e., "hold *paramount* the safety, health and welfare of the public," takes precedence.

Design engineers should also evaluate the defeatability of any devices or features important to the safety of the product or design. Often there are very real, and sometimes powerful, incentives to remove, bypass, or defeat safety devices, or otherwise negate the safe function of our designs. A list of these incentives might include: authoritative orders or instructions (someone instructing a user to ignore safety procedures); illness, handicap, or infirmity; customs or habits; fatigue; discomfort, pain, or irritation; anger, overconfidence, or complacency; being late or in a hurry; inconvenience; cost; improper design of the safety feature; and human nature, including recklessness and thrill-seeking (3).

While it may be impossible to design a safety feature that can never be defeated, some designs are more easily defeated than others are. Remember that safety features that are well executed and do not hinder or otherwise restrict the intended use of the product are less likely to be removed or defeated. Warnings can also be included to ensure that anyone removing a guard or safety device is fully aware of the risk being assumed by such removal.

TESTING AND EVALUATION

Testing and evaluation during the design and prototype stages should consider all modes of failure and whether a failure results in a hazard. Obviously, some "failures" have few, if any, adverse consequences. There are many types of failures: overload failures, fatigue failures, environmentally assisted fractures, corrosion or erosion failures, failures associated with high temperatures or irradiation, etc. Consideration of every reasonably foreseeable failure mode is necessary in the design of a reasonably safe product or structure. Such failures include, but are not limited to, broken components that become dangerous projectiles, malfunctions that allow a machine to behave in an unpredictable manner, and minor component failures that precipitate chain-reaction events.

A potential ethical dilemma sometimes arises during the testing phase of product development, or during construction of a large civil project. A design that looked great during the design phase may fail during testing, thus prompting a costly redesign and delayed production or project completion. Such a situation can result in pressure to cut corners in the design to recoup lost income or improve the delivery schedule. Occasionally there may even be pressure (from colleagues, managers, or even clients) to falsify test results, i.e., to "cover up" a material defect or design flaw. Professional engineers must be alert to this pressure, and remember their responsibilities to the public safety.

FIELD EXPERIENCE

Good design engineers will also gather and study field experience information for early identification of any unknown safety issues. Compiling field experience data from similar products or designs during the initial design evaluation, and promptly evaluating any field problems once the new design is implemented, are the signs of a conscientious engineer. Correction of such problems early in a production cycle, or during initial construction, is almost always easier and less costly than a large-scale recall and/or retrofit of designs already in the market. Avoid the tendency to think of a redesign as an admission of a defect. Even post-production changes are preferable to continued exposure of the public to a known hazard.

ENCOURAGING PROPER USE

Ideally, products should be designed for easy, safe use with as few instructions as necessary. If possible, controls and other features should follow established and/or intuitive patterns to minimize confusion and inadvertent operator error. Concise text and universal symbols should communicate the proper use of the product and convey the hazards of misuse. Such communications should be in view of the operator and attached to the product, if possible—not relegated to a manual that may never be read.

Instructions or warnings should educate the user about a hazard and the proper means for avoiding it. However, human nature almost guarantees that an instruction or warning will never be 100% effective as a protection strategy. Therefore, use warnings only as a last resort (for hazards that cannot be eliminated by design or guarded against) and as a supplement to guarding. Refer to standards on the design of warnings and instructions to use lessons learned by prior generations of engineers. In an effort to improve effectiveness and reduce confusion, the design elements of product warnings have been standardized with respect to (a) signal words and colors, (b) text content, (c) letter size and format, and (d) location of the warning.

Standard warnings use the signal words " DANGER," "WARNING," and "CAUTION," used in descending order according to the severity of the hazard. "DANGER" notices use the colors red and white, while "WARNING" and "CAUTION" notices use the colors yellow and white. Other important notices, such as maintenance information and information not related to personal injury hazards, should use the signal words "IMPORTANT" or "NOTICE," and the colors black and white.

The text of any warning should contain at least the following three elements to fully communicate the nature of the particular hazard: (a) a clear statement explaining *what* specific injury might occur; (b) a clear description of exactly *how* the injury may occur; and (c) a clear statement explaining the *precautions* a user must take to eliminate the risk of injury. The letter size and location of warnings should be determined based upon the anticipated location where users must become aware of the hazard. They must be able to read the warning in time to initiate any action required to prevent injury to themselves or others.

Figure 9-2 provides an example of safety signs designed in accordance with American National Standards Institute (ANSI) Standard Z535 (6,7). These symbols are used throughout industry and are readily recognized because of their standard design.

COMMUNICATION, COORDINATION, AND DOCUMENTATION

Professional engineers are bound by the NSPE *Code of Ethics* to communicate—to our clients, management, or other responsible parties—any safety or health concerns we identify. If our warnings are ignored, we should appeal to the next higher authority, hopefully without violating a client confidence. However, engineers should avoid the temptation to exaggerate. This advice applies to both sides of the safety argument. Engineers may be tempted to downplay safety concerns when the design of their product is called into question. Or, they may be tempted to exaggerate the safety risks when trying to convince management (or a client) to incorporate a safety feature into a product. Remember that the third fundamental canon in the NSPE *Code of Ethics* instructs us to "issue public statements only in an objective and truthful manner." The truth will usually be evident in time, and while exaggeration may sway a decision in a favorable direction now, the loss of credibility is almost impossible to recover.

Engineers should be aware of the limitations of their expertise, and get assistance in other areas of the design, if appropriate. Many product failures result from failure to communicate with management, or with other disciplines, during the design process. In large companies, especially, there's often a battle for company resources between various internal organizations. In large civil projects, the conflicts may be between different companies or contractors. Such conflicts for resources can inhibit communication between organizations that should be working closely together. If this happens, safety can suffer. Each organization may assume that the responsibility for correcting a safety problem lies with one of the other organizations. And none of them wants to use its resources to correct "someone else's problem."

Some generalized guidelines apply to most types of product design, especially with respect to the necessary elements of a manufacturer's product safety program that will reduce the risks of project hazards (3). The basic elements of such a program are applicable to almost any product, whether consumer or industrial. Professional engineers should encourage their companies to implement a formal safety program. Having such a program in place strengthens the position of an engineer defending the public safety against other conflicting priorities when the time comes.

Documentation of design procedures is an essential part of the engineering design of reasonably safe products. It forces us to follow the established procedures and reduces the chances that a step will be inadvertently skipped. Such documentation becomes especially important if our company is sued for negligence, since it will provide the substantiation that reasonable procedures were followed during the design process.

Figure 9-2. Example of Standardized Safety Signs. (Source: Seton Name Plate Co., Branford, Connecticut, Catalog BJ-6, p. LJ4, 1998.)

STANDARDS

Studying codes and standards, including building codes and fire codes, is a good way to understand the types of hazards others have identified and considered. Since its establishment by Congress in 1972, the Consumer Product Safety Commission has actively worked for the establishment of numerous standards, both mandatory and voluntary, for the design of safer consumer products and home furnishings. There are numerous standards in almost any other area you can imagine. In some cases, no standard is specifically applicable to the product in question, especially for new products on the cutting edge of technology. However, studying the standards of similar or related products is a good way to learn how certain design concepts are used to accomplish a safe design.

There are many different kinds of standards, as illustrated in Table 9-4. Each serves a different purpose, but all can be important in designing a safe and effective product. Standards dictated by law, such as *Occupational Safety And Health Standards* (29CFR 1910) and others codified in the *Code of Federal Regulations*, are obviously mandatory. Others, such as those developed and published by the American National Standards Institute (ANSI) in conjunction with various technical societies, are technically voluntary.

Industry-accepted voluntary standards are often adopted by public authorities, and made mandatory *by reference* in building codes and through other legislative means. Also, if a voluntary standard is widely accepted, it soon defines the "standard" for reasonable design, Failure to comply with it may then constitute defective design in the eyes of the legal system. Note that adherence to established standards is usually considered only as a minimum requirement, since products that meet all of the requirements of a standard can still be judged defective.

Table 9-4. Types of Engineering Standards

Definitional standards, which establish precise, uniformly excepted meanings and nomenclature for materials, apparatuses, processes, and the like.

Specification standards, which prescribe the use of specific materials, designs, production processes, performance levels, and testing methods.

Interchangeability standards, which dictate the dimensions of products or equipment; they promote the uniformity, simplification, and compatibility of products.

Classification standards, which allow competing products to be compared on the basis of their relative quality or performance grading.

Recommended practice standards, which describe the instructions for installation, use, or maintenance to ensure safe or satisfactory product performance.

Testing standards, which specify the apparatus, procedures, and testing conditions needed to determine whether a product meets requirements of an applicable standard.

Safety standards, which address the safety of a product.

Protection standards, which come into play when the design engineer determines that the consequences of a product failure are too severe or the probability of failure is too great. (An example of complying with protection standards would be to add a fragment shield to a lawnmower, turbine, or pressure vessel or a containment vessel for a hazardous chemical process.)

Source: Reproduced by permission of ISI Publications, Inc. from S Brown, ed., Forensic Engineering: Part I, An Introduction to the Investigation, Analysis, Reconstruction, Causality, Risk, Consequences, and Legal Aspects of the Failure of Engineered Products, 1995, ISBN 0-964-553600.

Professional engineers should work to establish standards if there are none, especially in areas that have a great influence on public safety. Few others will understand the interplay of the various design features on the safety of a product better than the design engineer will. The same logic applies to the tradeoffs of environmental or civil design. Of course, work within your company's rules regarding proprietary information or designs, since safety innovations may represent a significant advantage in the marketplace and may even be patentable. However, remember that improving the safety of your product industry will ultimately improve the bottom line of your company in most cases. Entire industries have been ruined by failure of manufacturers to evaluate and improve the safety of their products. Also, watch out for the tendency to use standards as a weapon against competitors in your industry, if you serve on a standards committee.

SUMMARY

In summary, design of reasonably safe products requires professional engineers to become educated about the legal system as it relates to product liability, so that we can protect the interests of our firm or clients. It also requires us to practice with a reasonable standard of care in our chosen field, both to protect the public and ourselves from negligent errors or omissions. We must familiarize ourselves with the tools of our profession and faithfully execute them.

STUDY QUESTIONS

9.1 Consider the following case from the NSPE Board of Ethical Review (NSPE Case No. 88-5):

Engineer A is employed by a computer manufacturing company. She was responsible for the design of certain computer equipment several years ago. She signed off on the drawings for the equipment at that time. Although Engineer A's design was properly prepared, the equipment manufacturing process was faulty and, as a result, the equipment became too costly and suffered mechanical breakdown. The manufacturing division made a number of recommended modifications to her design that it believed would help reduce costs in the manufacturing process. Engineer A's analysis of the manufacturing division's recommendations revealed that they would reduce the reliability of the product and greatly increase the downstream costs to the company through warranty claims. Engineer A's supervisor, who is not an engineer, asks Engineer A to sign off on the changes for the new computer equipment. There is nothing to suggest that the equipment would pose a danger to the public health and safety. Engineer A raises her concerns to her supervisor but nevertheless agrees to sign off on the changes without further protest. Did Engineer A fulfill her ethical obligation by signing off on the changes without further action?

9.2 In the above question, assume that Engineer A suspects a possible safety problem and that all other facts are as stated in the problem. Under these circumstances, would your conclusions be different? Why or why not?

9.3 Consider the following case from "You Be the Judge," a regular column appearing in *Engineering Times*, the newsletter of the National Society of Professional Engineers:

Duane Etright, PE, is employed by a software company and designs specialized software used in the operation of facilities affecting the public health and safety, such as air quality control and water quality control facilities. As a part of the design of a particular software system, Etright conducts extensive testing, and although the tests show that the software is safe to use under existing standards, Etright is aware of new draft standards that are about to be released by a standard-setting organization—standards that the new software might not meet. Testing is very costly and the company's clients are eager to move forward. The software company wants to satisfy its clients and protect its finances and employees' jobs, but at the same time wants to be sure that the software will be safe to use under the new standards. Tests proposed by Etright will likely result in a decision on whether to move forward with the use of the software. The tests are costly and will delay the use of the software by at least six months, which will put the company at a competitive disadvantage and cost it money. Also delaying implementation will cause the state public service commission utility rates to rise significantly. The company requests Enright's recommendation on the need for further software testing. What should Etright recommend?

REFERENCES

1. Webster's Ninth New Collegiate Dictionary, Merriam-Webster, 1988.
2. TA Hunter. Engineering Design for Safety. New York: McGraw-Hill, 1992.
3. S Brown, ed. Forensic Engineering: Part I, An Introduction to the Investigation, Analysis, Reconstruction, Causality, Risk, Consequences, and Legal Aspects of the Failure of Engineered Products. Humble, TX: ISI Publications, 1995.
4. RM Seiden, PE. Breaking Away: The Engineer's Guide to a Successful Consulting Practice. Englewood Cliffs, NJ: Prentice-Hall, 1987.
5. An Instructional Aid for Occupational Safety and Health in Mechanical Engineering Design. New York: American Society of Mechanical Engineers, 1984.
6. American National Standards Institute. American National Standard Criteria for Safety Symbols. Rosslyn, VA: National Electrical Manufacturers Association, ANSI Z535.3-1998.
7. American National Standards Institute. American National Standard Criteria for Product Safety Signs and Labels. Rosslyn, VA: National Electrical Manufacturers Association, ANSI Z535.4-1998.

10

Compensation and Employment Ethics

COMPENSATION ETHICS*

The subject of compensation is one of those topics that make engineers, and most "regular" people, uncomfortable. While engineering students know that engineers are generally well compensated, few know very much about how engineers are compensated in the real world. And even fewer are prepared for the ethical dilemmas often encountered in the realm of compensation.

There are multitudes of ways for engineers to be compensated for their talents. In industrial settings, compensation generally involves a base salary arrangement, sometimes with profit sharing and other incentive programs. Engineers may or may not be paid overtime pay for time in excess of a normal work week. And a normal work week might be defined as something other than 40 hours. Usually, salaried engineers and other professional employees are exempt from wage restrictions that tightly control overtime pay, benefits, work hours, etc., for hourly employees. In these settings, engineers typically log their time and expenses using some type of time voucher system, with the labor charges billed to specific projects, to an external client or internal account, or sometimes even to the U.S. government.

Concerns about Using Time Voucher Systems

One of the insidious features of this time voucher system is that engineering employees generally do not have much knowledge of the accounting system. They rarely, if ever, see the dollar amount invoiced to a client or project, and may have

----

* Section by Russell C. Lindsay, PE, PARC Engineering Associates, Asheville, North Carolina

little real experience with project budgets. Thus, an engineer may not think twice about charging time to Project A, even though the engineer was actually working on Project B. After all, the Boss told him to do it, and he still gets paid the same. What he may not realize is that Project B is an internally funded development program, while Project A is a government-funded technology improvement program. The internal project is over budget and the Boss is attempting to complete it using government funds. By charging his time to the wrong project, the engineer is defrauding the U.S. government. A similar scenario has been the source of large civil and criminal penalties against more than one large, multinational corporation, resulting in financial losses for the company and ruined careers of numerous executives. This type of improper accounting can occur in a small private-practice firm, too. If discovered, the reputation of the company can be permanently tarnished by the public scandal, adversely affecting the job security of employees. Engineering employees knowingly participating in such schemes may also be subject to criminal prosecution.

Engineers in Private Practice

Engineers in private practice may be compensated using an arrangement similar to that in the industrial world, i.e., a salary plus some provision for overtime pay and/or incentive programs. However, private practice engineers usually have much more intimate contact with clients than engineers in industry, and often have direct authority to influence how the firm is compensated. The firm's compensation may have a direct relationship, through a profit sharing or similar arrangement, to the engineer's individual compensation. Thus private practice engineers must be attentive to avoid the pitfalls of competitive bidding, conflict of interest, and other compensation issues, as discussed in the following paragraphs.

Retainers Ensure Objectivity

In forensic (legal) consulting, compensation is a key ethical concern in that field of practice. Financial incentives are probably the most common type of a conflict of interest for the forensic engineer. Thus, the forensic engineer must be vigilant to ensure that compensation policies and practices do not provide incentives for "interest-based" decisions and opinions. The use of a retainer for most assignments is an easy way to ensure that compensation for forensic assignments is not "contingent" on the outcome of a case, or on the final determinations of the engineer. In the event that a retainer arrangement is not used, there must be clear understanding between all parties that there is no contingency arrangement. Forensic consulting is discussed further in the next chapter.

Retainers and advance progress payments are also useful methods of compensation in other fields of consulting practice. These advance payments keep the financial status clear for all involved, and keep the project from going over budget without all parties being aware of it. They also keep the potential worries about payment from entering into the engineer's judgement process. Additional project expenditures can be recommended, if warranted, without fear of jeopardizing payment for services already rendered.

Competitive Bidding

Earlier versions of the NSPE *Code of Ethics* prohibited competitive bidding. However, the courts have ruled that this prohibition unlawfully interfered with the right of engineers to provide price information to their clients, with specific mention of protections under the Sherman Antitrust Act. Thus, engineers are allowed to participate in competitive bidding, but the Supreme Court clearly stated, in its decision of April 1978, "The Sherman Act does not *require* competitive bidding." Furthermore, the Supreme Court decision made it clear that engineers and firms may *individually* refuse to bid for engineering services, and clients are not *required* to seek bids for engineering services [emphasis added]. Many engineering societies, NSPE included, continue to advocate the qualifications-based selection (QBS) process in awarding contracts for engineering services.

Plan Stamping and Compensation

As you probably know, building code officials and other authorities require engineering approval, in the form of a professional engineer's seal, on many civil/structural, environmental, electrical, mechanical, or other major projects. In addition, engineers are often called upon to "approve" the safety and integrity of existing structures or designs. Many types of remediation work also require engineering approvals. This type of assignment may appear, at least to the layman, to be easy work. The detailed analysis behind the sealed plan or drawing is not readily apparent to most people, and thus it may appear that the engineer is simply being paid for "plan stamping."

In fact, some unscrupulous engineers do plan stamping without proper analysis or supervision of the design. They accept payment without fully understanding what they have approved, or in some cases approve obviously defective designs if the fee is high enough. This practice adversely affects the reputation of the entire engineering profession and is contrary to the NSPE *Code of Ethics* and the engineering licensing laws of the various states and territories.

Double Dipping

Sometimes engineers find themselves in a situation where they can be compensated by more than one party for services on the same project. Probably the most frequent occurrence of this situation is where an engineer or members of a firm serve in both a public and private capacity. For instance, an engineer may provide services to a developer while also serving on a board or commission that approves the development. Alternatively, an engineering firm might wish to bid on a contract with a governmental agency. If one of the principals of that firm also serves on the governing board of that agency, the apparent conflict is obvious.

Generally, this practice is unethical and is specifically forbidden in the various engineering codes of ethics. Only in rare cases, where all parties are fully aware of the compensation arrangements, and approve them, is this practice allowed.

Kickbacks and Commissions

Most of us would agree that bribes are unethical. But surely commissions are acceptable, right? And what about political contributions that are intended to influence the award of a contract? The NSPE *Code of Ethics* prohibits engineers from offering, giving, soliciting, or receiving, even indirectly, such political contributions. The prohibition extends to "gifts, or other valuable consideration" offered in order to secure work. Commissions are forbidden too, except to bona fide employees or marketing agencies. Commissions, or kickbacks, between engineers and vendors, or engineers and clients, are considered unethical. The reason for these prohibitions goes back to the conflict of interest consideration.

In the area of compensation, engineers must pay particular attention to avoiding any *appearance* of impropriety. Appearances can be even more important if the project is a highly visible public endeavor, as many engineering assignments are. To preserve our individual reputations, not to mention that of our profession, we must often pay attention to how things look, even if there is no real impropriety. For example, how does one decide if a gift constitutes a commission or kickback? Certainly, few people would feel compelled to specify vendor A's product over vendor B's simply because vendor A gave them a complimentary coffee mug with their name on it. Yet some major corporations require detailed accounting of even such insignificant gifts. The reason is simple—*appearances*! While such gifts probably involve no conflict of interest in the strictest sense, most of us would have trouble deciding where to draw the line. What level of gift constitutes an undue influence? Does a free engineering design, provided by an equipment supplier in exchange for specifying its product, constitute an undue and conflicting influence? The NSPE *Code of Ethics* says it does. Perhaps we should consider whether we could defend our acceptance of such a gift if challenged by the journalists on the TV news magazine, *60 Minutes*.

Moonlighting

What about the engineer who provides individual engineering services outside of normal employment? As with many compensation issues, the key element is disclosure and approval by all parties. Contractors, clients, and others may know an engineer employed by ABC Engineering through this employee relationship. While the engineer may be contracted to perform work as an individual, the engineer's reputation and that of the employer are closely related. Thus, while the employer gains no compensation from the engineer's moonlighting activities, the firm is exposed to the risk of negative publicity, and potentially even liability, if the engineer is negligent. With this in mind, it should be obvious that permission should be obtained from the firm before pursuing after-hours work. Likewise, clients who retain the individual engineer should be fully aware that work performed under these circumstances does not have the resources, financial and otherwise, of the firm behind it.

Moonlighting engineers may be tempted to bill their services at a significantly discounted rate relative to their employer's rate for their services. Obviously, this may be the primary reason clients would hire the individual rather than hiring the firm. However, for the protection of their clients, their employer, and themselves, moonlighting engineers should consider obtaining professional liability and other types of insurance coverage for their activities. This may significantly increase the compensation rate required to make moonlighting worthwhile.

Engineers as Employers

Many engineers eventually become managers, owners, or principals in a corporation or engineering firm. This position places them in a situation where they may make salary and compensation decisions regarding other engineers. Here, engineering managers or employers are bound by the principles of appropriate and adequate compensation for engineers, based on professional qualifications and not on other inducements. In other words, engineers are prohibited from accepting remuneration from an employee or employment agency for providing a job. One might think of this as qualifications-based selection (QBS) on an individual employee level.

EMPLOYER ETHICAL VALUES

Corporate Ethics Policies

It is incumbent upon engineering employers to conduct their business affairs in an honest and ethical manner. To this end, most employers have a formal ethics policy that spells out what is expected of its employees and officers. Many firms also have an ethics office or special administrative department that is charged with assuring that the company policy is followed.

Raytheon Corporation, a major manufacturing corporation, is typical and has a series of formal printed ethics policies covering the various aspects of its business. As an example, Raytheon booklet No. 10-1924 describes in very clear and understandable terms its "Standards of Ethical Conduct for Supplier Relationships" (1).

The introduction of this booklet reads in part:

> *If you're in a position to spend company money, or to influence spending, you're a tempting target. People who sell to us may try to influence you to give preference to their materials or services...Raytheon expects you to resist when anyone tries to ply you with gifts of favors. You must always base your buying decisions on*

competitive price, quality and delivery. The company expects you to have friendly relationships with suppliers. At the same time, you need to be open, honest, businesslike and completely ethical...This brochure points out the main elements of the standards of conduct the Company expects of you...Raytheon has a reputation of integrity to uphold. And so do you.

The Raytheon booklet goes on to specify that gifts, services, or other considerations from suppliers—other than advertising novelties of nominal value—must be refused. Even advertising novelties are to be refused if they have an apparent value of $10 or more. The policy also states that it applies to members of the employee's immediate family.

Next, the policy spells out that proprietary information must not be offered to suppliers and that, if a supplier divulges information which is proprietary to its business, the information must be held in confidence "...both for the supplier's sake and...for Raytheon's sake too."

Last, the booklet clearly states that all applicable laws must be observed in procurement relationships and that kickbacks of any type are prohibited and are illegal.

Raytheon distributes this booklet to both employees and suppliers and provides toll-free telephone numbers for reporting of unethical conduct to the Raytheon Ethics Compliance Office.

This booklet is only one of many Raytheon documents defining company ethics policies, but it is typical and it does illustrate the type of policies that are widely adopted by responsible industrial firms.

Government Ethics Policies

Most government agencies, whether local, state, or national, also have formal written ethics policies. A typical such policy is that of NCDOT, the North Carolina Department of Transportation (2). It is brief, only two pages, but it is reasonably comprehensive despite its brevity. The preamble reads:

The holding of public office by appointment or employment is a public trust. Independence and impartiality of public officials are essential to maintain the confidence of our citizens. The members of the Board of Transportation and the officers and staff of the North Carolina Department of Transportation have a duty to the people of North Carolina to uphold the public trust, prevent the occurrence of conflicts of interest, and to endeavor at all times to use their position for the public benefit. To this end, members of the board, officers, and staff of the Department of Transportation shall ensure that an atmosphere of ethical behavior is promoted and maintained at all times.

This short preamble sets a clear and unambiguous tone and indicates a high degree of ethical concern by this agency, a major employer of engineers in North Carolina.

The NCDOT document goes on to describe policies on conflicts of interest and gifts and favors. Officers and employees are urged to consult with the departmental general counsel or members of the North Carolina Attorney General's staff when ethical questions arise.

Failure to observe the NCDOT policy is grounds for disciplinary action including possible dismissal from employment.

Does this seem harsh or excessive? It shouldn't. It makes common sense and is fully consistent with the various engineering codes of ethics.

Students' Views of Employer Engineering Ethics

The Professional Engineers of North Carolina Ethics Steering Committee has developed a program to educate and promote awareness of ethics among engineers and engineering employers. The program includes several ethics case studies taken from past decisions of the NSPE Board of Ethical Review.

One of these case studies was used as a part of a final examination for engineering students in a course on "Introduction to Engineering". The circumstances of the case, as presented to the students, were as follows:

> *Smith, an unemployed graduate engineer who recently received certification as an Engineer-Intern from the North Carolina Board of Registration for Engineers and Land Surveyors, is seeking employment with a consulting firm. Smith is contacted by Engineer A, a principal with a large Charlotte area consulting firm. After a long discussion including such matters as working conditions, salary, benefits, etc., Engineer A offers and Smith accepts a position with the firm. Smith cancels several additional job interviews with other individuals.*
>
> *Two days later, in a meeting with other principals of the firm, it was agreed by the firm's management, including Engineer A, that the vacancy should be filled by an engineering technician, not a graduate engineer. A week and a half later Engineer A contacted Smith and rescinded the firm's offer.*
>
> *Did the actions of Engineer A in his relations with Smith constitute unethical conduct? Yes or no? Discuss your reasons.*

The responses of the students to this problem were interesting and varied. One student concluded that Engineer A's actions were unethical but that Smith was stupid in not getting the offer in writing (presumably so that he had legal proof of the offer) and also for canceling his other job interviews.

Another student did not consider Engineer A's actions to be unethical because it was a purely business decision. However he felt that Engineer A owed an apology to Smith.

A third student felt that Engineer A should have obtained verification that the offer was appropriate before extending it to Smith and also that, once the decision was made to rescind the offer, Smith should have been notified immediately. This student felt that Engineer A acted improperly but not necessarily unethically.

The remainder of the students was unanimous in their opinion that Engineer A's actions were unethical and gave similar reasons for their opinions. One student's words summarize the consensus opinion rather well:

> *The nature of Engineer A's discussion of the position, including salary and benefits would lead an individual to perceive any offer of employment as a valid one. If Engineer A was not in a position to make an offer, one should not have been made. Management is entitled to reconsider their decisions but this should have been relayed to Mr. Smith up front. Instead he was offered a position and he accepted. Reversing a legitimate offer for purely economic reasons is poor business conduct. Engineer A's implication that his offer was a final one can be considered deceptive and, therefore, unethical.*

The NSPE Board of Ethical Review in the actual case concluded, in part:

> *There can be no doubt that just as a prospective employee has an obligation to act in good faith with a potential employer, an engineer employer owes a duty to deal honestly, fairly and openly with a prospective engineer employee...*
>
> *While it certainly must be acknowledged that in difficult economic times employers of engineers must make difficult decisions that are frequently unpleasant and incur human cost, this fact should not in any way diminish the basic and fundamental obligation and responsibility of all engineers and employers of engineers to be mindful that such actions, if taken in a more careful and conscientious manner, could result in less hardship and embarrassment for all concerned...*
>
> *The actions of Engineer A, in his relations with Smith, constituted unethical conduct. Furthermore, the involvement of all the principals of the firm, in connection with the withdrawal of the employment offer, constituted unethical conduct.*

What would you have decided in this situation? Would you have agreed with the Board of Ethical Review and the majority of the engineering students?

Reaction to the Students' Views

The discussion above about students' views of engineering ethics was published in the September/October 1996 issue of "the Professional Engineer, the magazine of North Carolina Engineering" and prompted responses from some readers.

One response was particularly noteworthy. Thomas M. Stout, PE, saw the article during a visit to the offices of the California Society of Professional Engineers in Sacramento.

With his permission, his response is quoted verbatim:

> *In June 1947, while working on a master's degree at the University of Michigan, I had an interview with one of the big electrical manufacturing companies. Things seemed to go well on both sides; they sent me an offer letter (for about $260 per month), and I sent back an acceptance, saying I would report in September after the end of the summer term.*
>
> *Back came a letter saying they were glad I had accepted, but there was one hitch; they had offered me $20 per month more than they should have. No apology was offered, only the comment that these things happen in large companies. They closed by saying they looked forward to my arrival.*
>
> *I wrote back and said I didn't want to work for a company that didn't honor its offers and wouldn't be reporting in September. I ended up applying to another company and accepting a similar offer.*
>
> *I thought a large company could easily absorb $20 per month and, if it was a serious problem, could have delayed my first raise long enough to recoup the small amount of money involved. I would not have known about their "error", would have a better taste in my mouth, and would have had an entirely different career!*

How would you or your company have reacted under the circumstances if you were the one who extended the offer in the first place and later discovered the "error"? How would you have reacted if you were the engineer? Would you have reacted as Tom Stout did? I certainly hope so.

Conclusion

L. G. Lewis, Jr., PE, a member of the South Carolina Board of Registration for Professional Engineers and Land Surveyors, commented in an NSPE paper (3) that, "These various rules of professional conduct [referring to the engineering codes of ethics and to engineering licensing laws and regulations], and the disciplined enforcement of these rules, are important. They do not, however, contain the final emotional power of commitment. Perhaps the individual's (and the corporation's) push for maximization of wealth is the major obstacle to achieving higher ethical standards of ethical practice...."

In the same article, Lewis quoted Supreme Court Justice Sandra Day O'Connor, who said:

> *One distinguishing feature of any profession is that membership in that profession entails an ethical obligation to temper one's selfish pursuit of economic success by adhering to standards of conduct that could not be enforced either by legal fiat or through the discipline of the market. Both the special privileges incident to membership in the profession, and the advantages those privileges give in the necessary task of earning a living, are the means to a goal that transcends the accumulation of wealth. That goal is public service.*

Lewis concluded his paper with the words, "Ethical standards of practice have never been more besieged than they are today. Those with strong moral fiber, a dedication to professional integrity, and the ability to reason soundly must find the power to resist the attack. The leadership to nurture such power within the office environment is a responsibility not to be taken lightly by corporate management."

Strong employer ethics policies can go a long way toward accomplishing this objective.

STUDY QUESTIONS

10.1 Engineer Cy T. Work, PE, is retained by Farms to Homes, Inc. to perform preliminary site studies of a large tract of land under consideration for purchase and development. Specifically, Work's analysis of the cost and feasibility of roads, utilities, and drainage structures is a significant factor in the decision not to purchase the tract. Subsequently, Work is asked to perform similar services for another potential buyer. Is it ethical for Work to accept this subsequent assignment to perform essentially identical services for another client? Would it be acceptable if he obtained the permission of Farms to Homes, Inc. for release of his report to them? What other solutions to this dilemma can you think of?

10.2 An engineering firm is retained by the airport authority to design lighting systems for a new passenger terminal, parking garage, and taxiway. Six months after the project is completed, the local newspaper reports that the owner of the engineering firm was treated to an expense-paid trip to Bermuda, courtesy of the manufacturer of the lighting equipment he specified. Would you be suspicious of his motives in specifying that particular lighting manufacturer's products? Even if the gift was not prearranged, do you think his judgement will be affected on future projects?

10.3 Consider the following case from the "You be the Judge" column of *Engineering Times*:

Manny Hatz, PE, a licensed professional engineer and land surveyor, works 35 hours per week on a flex-time basis for a state governmental agency. In addition, Hatz is associated with XYZ Engineering and Surveying firm as the PE in charge under the state's certificate of authorization requirement.

Hatz provides about 20 hours per week supervising engineering services at the firm, plus an additional 12 hours of work on the weekends, and is available for consultation 24 hours per day. XYZ grants Hatz a 10% share of the stock in the firm, and as compensation for his engineering services, Hatz will receive 5% of the gross billings for engineering work for which the seal of a licensed engineer in responsible charge of engineering is required. The agreement is contingent on the understanding that if any one of the three principals of XYZ becomes licensed as a PE in the state, the agreement will become void and the 10% stock will be returned to XYZ.

Both the state governmental agency and the engineering firm are aware of Hatz's activities as a dual employee and do not object to these activities.

What do you think? Is it ethical for Hatz to be associated with XYZ Engineering and Surveying in the manner described?

10.4 Consider the following case from the NSPE Board of Ethical Review (NSPE Case No. 77-12):

Mary Smith, PE, a consulting engineer who practices primarily in the field of industrial product design for clients, is requested by the XYZ Manufacturing Company to review an amplifier design developed by the company. The company has not been successful in producing an acceptable product and is under pressure to deliver a final model to a customer within three months. Smith spends a few days reviewing the XYZ design and makes several recommendations to improve the product. She is paid her usual per diem fee, as earlier agreed upon. However, XYZ advises Smith that it will need her further assistance for the product in order to make it fully acceptable and proposes to retain her for further services on a basis that she will be paid a fee for the additional service only if the amplifier, as a result of her assistance, will meet the company's requirements. During this period XYZ will pay Smith her out-of-pocket costs, e.g., travel, lodging, computer time, etc.

Would it be ethical for Smith to enter into a contract arrangement as described?

REFERENCES

1. Standards of Ethical Conduct for Supplier Relationships. Form 10-1924. Concord, MA: Raytheon Corp., 1995.
2. Ethics Policy for Officers and Employees. Raleigh, NC: North Carolina Department of Transportation, 1993.

3. LG Lewis. The Cultivation of Professional Ethics. National Society of Professional Engineers. Undated. <http://www.nspe.org/eh1-lew.htm>

11

Ethics in Forensic (Legal) Consulting

CAREERS IN FORENSIC ENGINEERING*

The past few years have seen a dramatic increase in the use of all types of professionals as expert witnesses in litigation, especially civil litigation. And engineers of all disciplines are now often called upon to testify in depositions, arbitration hearings, trials, and other forms of dispute resolution. The demand for such services is so great that many engineers devote much, if not all, of their professional time to such endeavors. These engineers have come to be called forensic engineers, regardless of their technical specialty (mechanical, electrical, civil, environmental, etc.), and they are very likely also to be licensed as Professional Engineers.

The primary ethical dilemma for engineers in forensic practice stems from the fact that clients are often an advocate for one party in the dispute, but the forensic engineer is required to be objective, impartial, and dispassionate. Not all of the individuals who have entered this field are up to the challenge. And of course, as in any profession, the actions of a few unqualified, disreputable, or even misguided, individuals can reflect on the perceived ethical standards of an entire profession.

The growth of this field of practice, and the engineer's need for guidance, was recognized some time ago by the Professional Engineers in Private Practice (PEPP) division of the National Society of Professional Engineers (NSPE). They charged the Forensic Engineering Committee to study the issue and revise an existing NSPE/PEPP document to provide guidance to engineers in forensic practice. The result was the NSPE publication *Guidelines for the PE as a Forensic Engineer* (1). Every engineer involved in forensic work should read this document. It very concisely spells out the aspects of forensic work that make it different from

----

* Section by Russell C. Lindsay, PE, PARC Engineering Associates, Asheville, North Carolina

other engineering assignments and the professional practices that can help engineers work ethically in this field.

Another informative document for engineers involved in forensic practice is one entitled *Recommended Practices for Design Professionals Engaged as Experts in the Resolution of Construction Industry Disputes,* published by the Interprofessional Council on Environmental Design (ICED) and distributed by the Association of Soil and Foundation Engineers (2). The ICED document is also reprinted in another informative text on the subject, *Forensic Engineering* by Kenneth L. Carper (3). The practices spelled out in the ICED document echo the guidelines of the NSPE document, lending additional weight to the common obligations called out in both.

There are several main areas of ethical concern in the practice of forensic engineering, following the general pattern of all engineering, but with some wrinkles specific to forensic practice. Working only in one's areas of competence is obviously an issue, as is compensation and attention to potential conflicts of interest. Engineers in forensic practice will find that there is almost a constant tension between several of the Fundamental Canons of the NSPE *Code of Ethics*.

Competence

One frequent problem with forensic assignments is that the nature of the assignment is not always clear at the outset. Because of the largely investigative nature of forensic work, the character of the job can change radically as evidence and additional information becomes available. What at first appears to be an electrical failure may turn out to be a mechanical failure, and vice versa. Or an apparent structural design failure may actually be a corrosion/materials problem. While most engineers are cross-trained in other disciplines to some degree, forensic work requires close attention to deciding when an assignment has drifted too far outside one's primary areas of expertise. The client (often an attorney or insurance claims adjuster) will often apply pressure to "finish the job," because he wants to avoid the additional expense of hiring another specialist. A professional engineer should resist the urge to practice outside his chosen specialty, and be prompt and forthright in notifying clients as soon as the assignment turns in an unanticipated direction.

Forensic engineers must be careful not to lean too heavily on court rules of evidence that allow expert testimony when it "can assist the trier of fact to draw certain inferences from facts because the expert is better qualified" than the jury to interpret the facts. Under this definition, most engineers are "better qualified" than most jury members on a wide range of topics. However, a client will not be well served by an engineer that is not trained and experienced in the specialized discipline at issue. Limited experience and/or training increases the chance for error, either by overlooking a critical fact or by failing to fully understand a complex issue. Also, if the opposing counsel calls a more experienced and/or highly trained specialist engineer as a witness, the jury will be instructed to weigh the training and experience of the "experts" as part of their deliberations. Thus, while one engineer may arrive at the same conclusion as a more specialized engineer, his or her opinion will be given less "weight" in the eyes of the court if the engineer is not as qualified in that area. In this instance the client will be poorly served.

Most forensic clients are astute enough to realize that not every case can be "solved" by an engineering (or other professional) expert. However, they are also clearly in an advocate's role and may be oblivious to the technical weaknesses of their case. That is one of the reasons they need technical assistance from engineers. Remember that the most helpful service an engineer can provide to some clients may be to identify the weaknesses in their case as soon as possible, so they can settle it before wasting too much time and money (both theirs and that of the courts). Thus, once an engineer determines that his or her opinions are adverse to the client's interests, the engineer should tell the client so in no uncertain terms.

Compensation

Compensation arrangements should be easy, but they rarely are. The NSPE guidelines remind us that the *Code of Ethics* prohibits engineers from working on a contingency fee basis "under circumstances in which their judgement may be compromised." Thus, the financial arrangements must be such that payment of fees for service is not contingent on the outcome of any dispute resolution proceeding, trial, or settlement. This concept applies especially to the practice of "champerty," in which someone subsidizes the cost of a lawsuit, either directly or indirectly, in exchange for a share of the proceeds.

Ethical practice requires establishing a fee schedule, agreed to by both engineer and client, before accepting the assignment. Lump sum or fixed price arrangements are discouraged, since legal proceedings often involve delays and extensions beyond that which can be reasonably estimated at the outset. Providing the client with a nonbinding estimate of the anticipated costs of forensic services at the outset is reasonable, and often a good idea, but a per diem or hourly rate is usually the best type of compensation arrangement. An advance retainer is a good method for avoiding the appearance of a contingency arrangement. The amount of such a retainer should be based upon the anticipated scope of the assignment. It is preferable to maintain a consistent and uniform fee schedule for all clients and all assignments of similar nature. Most forensic engineers charge more for forensic assignments than for other engineering work, reflecting the greater demands of forensic work.

Since one of an engineer's duties is to act as a faithful agent or trustee of the client, regular and accurate invoicing of costs is required. This practice keeps the client informed of the cost of the assignment so that he has the information needed to decide on settlement terms, and it also maintains the credibility of the engineering expert. If the engineer has been working on the assignment for six months and has still not been paid (either because he never billed for his time or the client is delinquent), there is an appearance of impropriety even if the engineer and client have both agreed that the fee is not contingent on the outcome. Such appearances of impropriety can have an adverse effect on both the client's case and the reputation of the engineer.

Conflicts of Interest

There are many situations where a forensic engineer can encounter a potential conflict of interest. The most common is probably a situation in which the engineer is working for attorney A in one dispute and attorney B in another. The fact that this situation occurs should not be surprising, since all of the parties, and thus all of the attorneys, in a legal dispute are not always known at the outset. Such apparent conflicts *may* be acceptable (provided that both attorneys are made aware of it, and approve) since the attorney is an intermediary client in most cases. A less common situation is one where the conflict involves the "ultimate client," i.e., an actual party involved in a dispute. Generally, this type of conflict is unacceptable. Good practice requires maintaining some type of database of current assignments so that new assignments can be screened for potential conflicts before any confidential or privileged information is exchanged.

A forensic engineer may also be called upon to evaluate the technical competence of one of his own engineering peers, e.g., on some type of malpractice charge. This situation can feel like a conflict of interest situation because of the strong sense of ownership most engineers feel in their profession. Some engineers might feel inclined to protect the profession by refraining from criticism of other engineers in such instances. Others might take the opposite approach, holding the position that any engineer who disagrees with them must be incompetent. Upon reflection, however, it should be obvious that an engineer is obligated to participate in honest proceedings that serve to discourage unprofessional or incompetent engineering practice. A professional engineer is likewise obligated to protect the reputations of all parties until any investigation and analysis is complete. Remember that differences of opinions do not necessarily imply incompetence or unethical practice. While many aspects of engineering are based on strict mathematical or physical laws, a certain degree of subjectivity is always involved in engineering decisions and/or opinions. Also, remember that engineers are not expected to be infallible, but they are expected to perform their duties to a reasonable "standard of care," defined as "that degree of care which a reasonably prudent person should exercise in same or similar circumstances."

Other Pitfalls

There are many other potential pitfalls in the business of forensic engineering, most of them related to the unfamiliar workings of the legal system. Some of these pitfalls are detailed in a paper by Hunter (4). He describes the problems of the last minute assignment from an attorney on the eve of trial, in which the engineer is provided very limited opportunity (and time) to study the facts of the case and arrive at an opinion. Good judgment and a strong sense of ethics are needed to avoid being caught in an unworkable position in such an assignment. In many such cases the best approach is simply to decline the assignment. The time constraints of the situation, created by the attorney's failure to involve a technical consultant early in the process, results in pressure on the engineer to arrive at a conclusion extrapolated from too few facts. While in some cases a reasonable conclusion can be reached, a great deal of caution is warranted in accepting such cases given the loss of credibility that might result.

Another uncomfortable situation identified by Hunter is the dilemma of the client attorney who fails to disclose the existence of conflicting witness testimony. This places the engineer in the uncomfortable position of finding out about critical evidence from the opposing attorney, often during cross-examination in a deposition or on the witness stand. At best, this situation makes the engineer appear uninformed. At worst, the engineer has failed to consider all available evidence, and thus the engineer's opinion is biased. The chances of encountering this situation can be minimized by insisting that clients provide *all* relevant documents, testimony, and other evidence for review prior to formulating an opinion. Some clients may still withhold information, but after a while most forensic engineers develop a sixth sense regarding this practice.

Most difficult situations can be avoided by learning how the forensic engineer fits into the legal system, and paying close attention to the potential traps and pitfalls before they occur. Remember that in forensic engineering assignments the client is usually an advocate, but the forensic engineer must remain impartial. Honesty and integrity must be utmost in the mind of a forensic engineer, and even the *appearance* of impropriety must be avoided. The credibility of the entire profession, not just the individual, is at stake each time an engineer becomes an "expert" witness.

JUNK ENGINEERING

What is Junk Engineering?*

The derogatory term *junk engineering*, sometimes used interchangeably with the term *junk science*, refers to the practice by some engineers of stating opinions or conclusions not based in sound engineering or scientific principles. Often, the engineer stating such opinions is practicing outside his area of competence and thus may be unaware that his conclusions are unsound. Some disreputable engineers may also be willing to state engineering opinions or conclusions in situations where there is little basis or foundation for any conclusion. Such practices reflect poorly on the engineering profession, degrading the public's confidence in the ethical standards of engineers. An ethical and professional engineer should be fa-

----

* Section by Russell C. Lindsay, PE, PARC Engineering Associates, Asheville, North Carolina

miliar with the scientific and engineering principles underlying any opinions or conclusions, and with the assumptions inherent in any calculations performed.

In an effort to discourage junk engineering or junk science, the rules of evidence generally require that technical experts submit reports (or give testimony by affidavit or deposition) detailing not only the expert's opinions and conclusions, but the assumptions, facts, and references upon which their analysis is based, including any formulas or special calculation procedures. This requirement is designed to prevent technical experts from testifying to unsupported or spurious opinions, i.e., those without firm basis in accepted engineering or scientific principles. Unfortunately this requirement is not always successful.

Is Peer Review Needed?

In a March 6, 1998 *Knight Ridder/Tribune* news article, Dr David M. Priver talked about actions of the American Medical Association to curb junk science in the courtroom. According to the article, the AMA now considers expert witness testimony on the part of physicians as part of practicing medicine. Such testimony is now subject to peer review—evaluation by other qualified physicians—in the same manner that medical procedures are subject to peer review. If the review finds that a complaint about expert testimony has merit, the physician who offered the testimony can be disciplined and possibly lose his/her license to practice medicine.

Why did the AMA find peer review necessary? Never in the past has the peer review system been applied to testimony. Why then is it being done now?

The reason is the AMA concluded that action of this nature is necessary to protect the credibility of the medical profession and to assure that patients receive high quality medical care.

The medical profession has experienced problems with a small group of physicians who, in the AMA's judgement, practice junk science. These physicians allegedly create scientifically insupportable and somewhat fanciful stories in support of plaintiffs' lawsuits, then receive pay from plaintiffs' attorneys. Dr. Priver says, "Many are little more than today's equivalent of the hired guns of the Old West—ready to say anything on the witness stand for the highest bidder."

In the article, Dr. Priver gave the example of Merrell Pharmaceuticals withdrawing the drug Benedectin from the market in 1983. The drug, according to Dr. Priver, was a proven medication for nausea and vomiting during early pregnancy. Although millions of women successfully received the drug and benefited from using it, a few women gave birth to deformed babies.

Aggressive attorneys using a few physician "experts" asserted a connection with the use of the drug and filed lawsuits despite many studies reported in reputable medical journals that the use of the drug did not cause the defects.

Merrell vigorously defended itself and won almost all of the cases. The legal costs, however, were so massive that the company pulled the drug from the market. Since then, according to Dr. Priver, the incidence of severe nausea and vomiting during pregnancy has more than doubled in the United States, while these women and their unborn babies have been denied effective treatment.

Dr. Priver gave other examples of "junk science" and concluded that the physicians who engage in these practices are abusing "...their positions of trust by testifying dishonestly..." but, in the past, were never held accountable. It is this inequity which the AMA is attempting to correct.

The case discussed by Dr. Priver eventually precipitated a U.S. Supreme Court ruling (Daubert vs. Merrell Dow Pharmaceuticals, Inc.) which clarified Federal rules of civil procedure.

The ruling now requires that scientific testimony must be "relevant and reliable." Such testimony must be based on "scientific method or procedure" and comprise more than "subjective belief or unsupported speculation." A key question is whether the theory or technique can be, and has been, tested. A second consideration is whether the theory or technique has been subject to peer review and publication. Publication is not required for admissibility, but the value of peer review is a consideration in determining admissibility.

Well, what does this have to do with engineering? It has everything to do with our profession because junk science is not a problem unique to medicine. We have many highly qualified forensic engineers who do an outstanding job of analyzing engineering problems and testifying in court in a truthful and professional manner when asked to review a case. Nonetheless, we have our junk engineering practitioners. Often they are so-called experts on some facet of technology. They usually aren't registered Professional Engineers but rather are "experts" dug up by an aggressive attorney who gets them to agree to present a viewpoint on a technical matter, whether or not that testimony is supportable with scientific and engineering fact, not conjecture or supposition. These "experts" are often discounted by testimony of Professional Engineers, but juries are unpredictable and sometimes cases are won by attorneys who use the junk engineering experts. Both plaintiff attorneys and defense attorneys, along with their "experts," are guilty of this practice.

Perhaps it is time for the engineering profession to follow the lead of the medical profession. What if expert testimony in engineering matters was legally considered to be the practice of professional engineering and was subject to peer review? The bogus, nonregistered "experts" would face penalties for practicing engineering without a license and the rare PE who, in effect, sells his/her seal in the courtroom would risk losing his/her license.

Another View of Junk Engineering

The preceding discussion suggesting that expert testimony in engineering matters be legally considered to be the practice of professional engineering was previously published in the November/December 1998 issue of *the Professional Engineer* (5). After the article was published, Eugene F. Rash, PE and attorney-at-law, commented as follows:

"The details of one particular case in which I helped in defending the designer are an excellent example. It was a South Carolina design-build project in which the general contractor built the structure but did not install the manufacturing equipment. The general contractor hired an outside architect-engineering (A/E) firm to design the building. A piece of owner-installed equipment that was not designed or even reviewed by the A/E exploded approximately 14 years after construction. This equipment was not in the construction contract, and therefore was not in the A/E's contract. There was no evidence that the general contractor, its subcontractors or its A/E ever touched the equipment. The equipment was stored on site during the punch list period, but there was no evidence that it had been installed prior to the final inspection of the punch list work.

"The decedent's estate sued the general contractor, its mechanical/plumbing subcontractor and the A/E based on an expert's opinion that both the general, the mechanical/plumbing subcontractor and the A/E had an implied duty in the industry to review the installation of the owner-installed equipment and advise the owner of any necessary safety measures omitted by the manufacturer. The expert further opined that the general contractor, the subcontractor and the A/E were all grossly negligent in failing to do so. The allegation of gross negligence was necessary to overcome the statute of repose in South Carolina and it would allow a jury to award punitive damages. The expert stuck with his position in spite of the plant engineer's deposition testimony that he installed the equipment on behalf of his employer (the owner), that the owner considered the equipment a trade secret, that it was designed by a separate entity under contract with the owner, and that the assistance of the general contractor, its subcontractor or the A/E was never requested.

"The plaintiff's expert was a PE in South Carolina and several other states, and had worked for a general contractor at some point. After a year of discovery, one of the better judges in the state was assigned to hear our motion for summary judgement and he granted the motion. Some judges would have relied on the expert's affidavit, denied our motion and sent the general contractor, its subcontractor, and the A/E to trial. The jury would have seen pictures of this fellow burned from head to toe. He lived several days after the accident, so there was considerable pain and suffering. His four young kids would have been sitting in court, probably crying from time to time.

"The owner was not a defendant because it was protected under the worker's compensation statutes. The manufacturer of the equipment was out of business, so it was not a defendant. A jury verdict could have been in the millions simply out of sympathy if they believed the plaintiff's PE. I believe it would have taken hundreds of thousands for the A/E to settle prior to trial had we not gotten summary judgment."

After receiving Mr. Rash's comments, the affidavit (6) and deposition (7,8) of the expert in this case were examined. The names of the parties involved in the case have been deleted but the quotes which follow are otherwise unchanged from the court documents.

The affidavit submitted by the expert states with respect to the defendants that they "...knew, or should have known...that the start-up and operation of a pressure vessel, such as Melter #1, without a safety relief valve was inherently and obviously dangerous" and "Failure by [the defendants] to provide basic safety features to protect construction workers and plant operators from a known and life threatening danger, is grossly negligent."

Subsequently, in a deposition (7) the expert was asked, "And if the owner took the position that the equipment that's going in this room is highly sensitive, very proprietary—'We don't want anybody to see it. We don't want anybody to touch it'—then the owner would be justified in telling [a defendant], 'We're only going to give you loads and information without letting you actually have hands on the equipment'?" His counsel objected. The question continued, "If in fact, that was [the owner's] position?" He replied "Yeah. Gosh...it would be unusual for that process at that time to be proprietary, but given your question, the answer to your question is yes."

Further on in the deposition, the expert was asked, "...you don't have any information that suggests that [a defendant] had a duty to design this melter?" He answered, "Oh, absolutely not. No."

In a later continuation of the deposition (8) the expert said, "...my opinion is that [a defendant] designed the connection that ran from the boiler to melter number 1." He was then asked, "And you haven't seen any drawing to that effect?" He replied, "No sir. That's correct."

Readers can draw their own conclusions about this testimony.

SUMMARY

Junk engineering is a problem of growing magnitude and is not limited to plaintiffs' experts. It occurs on the defendants' side also, although probably not as frequently. Nevertheless it is a serious ethical problem which the engineering profession must address.

The field of forensic engineering is an exciting one for many engineers. In addition to technical competence, this field requires good business and social skills to navigate the often unfamiliar waters of the legal system. Successful long-term professional practice requires unswerving adherence to ethical standards and close attention to compensation, technical competence, and other sensitive issues.

STUDY QUESTIONS

11.1 What is the typical legal definition of an expert?

11.2 Who are the typical clients of forensic engineers?

11.3 Does the ethical practice of forensic engineering allow contingent fee arrangements?

11.4 What is the preferred compensation arrangement for forensic engineers?

11.5 Why does a last-minute assignment on the eve of trial present a potential ethical problem?

11.6 What is "junk engineering," and how does the court system strive to exclude it from the legal system?

11.7 The following case appeared in the "You Be the Judge" column of *Engineering Times*, the newsletter of the National Society of Professional Engineers:

M. N. Nant, PE, is retained by Attorney James Smith to serve as an expert witness in a case under litigation. As part of the services required, Smith asks Nant to provide "additional assistance" in drafting cross-examination questions for use with other witnesses, analyzing arguments of the opposing counsel, assisting Smith in case strategy for trial, evaluating the credibility of lay witnesses, and critiquing Smith's opening and closing arguments. Nant has no legal training, but has ten years of experience in forensic engineering that includes serving as an expert witness for both the plaintiff and the defense bar.

What do you think? Would it be ethical for Nant to provide all of the services requested by Smith?

11.8 In the above question, assume that Nant held a law degree in addition to his engineering degree. Under this circumstance, would your answer to question 11.7 be different? Why or why not?

11.9 The following case also comes from *Engineering Times'* "You Be The Judge":

Samuel Smith is killed as a result of an allegedly defective product. Smith's widow can't afford to hire an expert to investigate the technical issues relating to the allegedly defective product. Instead she contacts A. Wise, PE, a product design specialist who frequently provides forensic engineering and related services. After a long discussion, Mrs. Smith asks Wise to act as her agent in the event she decides to sue the manufacturer. She requests that Wise advise her on selecting and retaining an attorney and overseeing the attorney's general approach to the case. Wise would not serve as an expert witness as part of the legal action. Since Mrs. Smith has limited resources, Wise, after considerable contemplation, agrees to perform the services on a contingency fee basis, in which he would receive a percentage of the settlement award. Was it ethical for Wise to agree to provide the services on a contingency fee basis?

11.10 In the above case, assume that Wise also agreed to give expert testimony at the trial. Would this fact change your answer to question 11.9? Why or why not?

11.11 What is your opinion in the case discussed on pages 128-129. Was the expert practicing junk engineering? Support your conclusion with appropriate citations from the NSPE *Code of Ethics*.

REFERENCES

1. Guidelines for the PE as a Forensic Engineer. NSPE Publication 1944. Alexandria, VA: National Society of Professional Engineers, 1985.
2. Interprofessional Council on Engineering Design. Recommended Practices for Design Professionals Engaged as Experts in the Resolution of Construction Industry Disputes. Silver Spring, MD: Association of Soil and Foundation Engineers, 1988.
3. KL Carper, ed. Forensic Engineering. New York: Elsevier, 1989.
4. TA Hunter. Some Hazards of Being an Expert Witness. ASME Paper 92-WA/DE-9. New York: American Society of Mechanical Engineers, 1992.
5. KK Humphreys. The Ethics Corner - Junk Science and Engineering. the Professional Engineer: The Magazine of North Carolina Engineering 31(3):18-19,1998.
6. Affidavit of [an Expert Witness], C.A. No.: 95-CP-23-2016 and C.A. No.: 95-CP-23-2017, In the Court of Common Pleas, State of South Carolina, County of Greenville, 5 September 1997, pp. 7-8.
7. Deposition of [an Expert Witness], C.A. No.: 95-CP-23-2016 and C.A. No.: 95-CP-23-2017, In the Court of Common Pleas, State of South Carolina, County of Greenville, 13 November 1997, pp. 229-230, 249.
8. Continuation of Deposition of [an Expert Witness], C.A. No.: 95-CP-23-2016 and C.A. No.: 95-CP-23-2017, In the Court of Common Pleas, State of South Carolina, County of Greenville, 9 December 1997, pp. 229-230, 249.

12

Whistleblowing

WHAT IS WHISTLEBLOWING?

The word *whistleblowing* carries negative connotations in the mind of many. As children, we were often admonished not to tattle on someone. If we did, we were severely chastised by our peers. "Telling tales" or informing parents or any type of authority that someone had done something wrong was simply not accepted behavior.

That childhood experience often sticks with us into adulthood and we are loath to report misdeeds. There is also the human reaction "not to get involved." We fear the consequences and personal inconvenience of becoming enmeshed in controversy or legal actions. Nevertheless, we must resist the urge not to whistleblow when we become aware of illegal, harmful, or unethical acts. In fact, under the laws of the various states and the various engineering codes of ethics, engineers are required to report anything of which they are aware which might negatively affect the "safety, health and welfare of the public" or which is fraudulent or illegal. The sixth Fundamental Canon of the NSPE *Code of Ethics* states that engineers shall "*...conduct themselves honorably, responsibly, ethically, and lawfully so as to enhance the honor, reputation and usefulness of the profession.*" That admonition alone requires that engineers not "hide their heads in the sand" when they become aware of improper practices—they must take appropriate action to assure that the situation is rectified.

Whistleblowing has been defined as "...exposing the misdeeds of others in organizations in an attempt to preserve ethical standards and protect against wasteful, harmful, or illegal acts." (1). In other words, whistleblowing is "going public", bringing the situation to light in a public manner. Whistleblowing is "...disclosure when the whistleblower steps outside of approved organizational channels to reveal a significant moral problem." (2).

WHEN SHOULD WE BLOW THE WHISTLE?

This question is not easy to answer. Clearly one should not blow the whistle when the intent of doing so is one of personal gain. The little boy who exclaims, "Billy took an extra piece of pie," is probably not concerned with ethics. His motivation is probably more of wanting the piece of pie for himself.

The following checklist offers some guidelines in determining the moral status of whistleblowing (1). If the first three conditions exist, whistleblowing should be considered. If all five exist, whistleblowing is obligatory.

1. The product or policy could do serious and considerable harm to the general public, the user, or innocent bystanders.
2. The problem should be reported to the engineer's immediate superior and the engineer's moral concerns made known.
3. If the superior does nothing effective about the concern or complaint, the engineer should exhaust the internal procedures and possibilities with the firm (assuming that an immediate public danger does not exist).
4. Accessible, documented evidence must be available that would convince a reasonable, impartial observer that the engineer's view of the situation is correct, and that the product or practice poses a serious and likely danger to the public or to the user of the product.
5. The engineer must have good reasons to believe that by going public the necessary changes will be brought about.

In other words, the engineer must determine whether there is wrongdoing, whether the situation can be rectified, and whether it is within the engineer's power to correct the problem.

Bright, McCall and Iyoob (1) list seven steps in determining if whistleblowing is the appropriate course of action.

1. Recognize the ethical dilemma.
2. Get the facts.
3. Identify the options for whistleblowing.
4. Test the options.
   - Is it legal?
   - Is it right?
   - Is it beneficial?
5. Decide which option to follow.
6. Double check your decision by asking, "How would I feel if my family finds out about my decision?" and "How would I feel if my decision is printed in the local newspaper?"
7. If all of the prior steps show that the *Code of Ethics* has been, or will be, violated, *blow the whistle*!

THE CONSEQUENCES OF WHISTLEBLOWING

The biggest factor that discourages engineers and others from whistleblowing is *retaliation*. Retaliation is an almost inevitable consequence of whistleblowing. Weil (2) said:

> When whistleblowers make their disclosures, others stand accused. People who feel accused or allied to those accused tend to hit back. That deflects attention from the accused. Retaliation is very damaging. That is why advice to whistleblowers includes the recommendation that the whistleblower resign in advance or in connection with blowing the whistle....
>
> The predictability of retaliation points to the moral complexities of whistleblowing. Both the accusation and the retaliation produce injury. Accusations not only threaten the careers of those on whom the whistle is blown; they also disrupt collegial relationships and other informal relations and networks. **Nevertheless, the whistleblowing may be morally justifiable when the moral wrong is very serious** [emphasis added]. Retaliation injures whistleblowers, spouses, and those who depend on whistleblower's earnings. One could argue that the potential whistleblower is obligated to include family members in the process of reaching a decision about blowing the whistle.

The whistleblower can expect severe consequences. However, the whistleblowing engineer must ask if the personal consequences are worth the effect upon the lives of others. Silence is a very poor choice when the safety, and possibly the lives, of others hang in the balance. It has been said that "physicians can bury their mistakes"—an engineer can't. They are there for all to see.

THE CHALLENGER DISASTER

The ethical engineer must make the hard choice about whistleblowing, as Roger Boisjoly did when the Challenger exploded.

On January 28, 1986, seven astronauts were killed when the space shuttle Challenger exploded just 73 seconds into the flight. The O-rings on one of the solid rocket boosters failed to seat properly allowing hot combustion gases to leak from the side of the booster and cause a flame which burned through the external fuel tank. Subsequent investigations attributed the O-ring failure to several factors, including faulty design of the solid rocket boosters, insufficient low-temperature testing of the O-ring material and the joints that the O-ring sealed, and lack of proper communication between different levels of National Aeronautics and Space Administration (NASA) management (3).

Background

Texas A&M University (TAMU) (3) summarized the background of the disaster as follows:

"NASA managers were anxious to launch the Challenger for several reasons, including economic considerations, political pressures, and scheduling backlogs. Unforeseen competition from the European Space Agency put NASA in a position where it would have to fly the shuttle dependably on a very ambitious schedule in order to prove the Space Transportation System's cost effectiveness and potential for commercialization. This prompted NASA to schedule a record number of missions in 1986 to make a case for its budget requests. The shuttle mission just prior to the Challenger had been delayed a record number of times due to inclement weather and mechanical factors. NASA wanted to launch the Challenger without any delays so the launch pad could be refurbished in time for the next mission....There was probably also pressure to launch Challenger so it could be in space when President Reagan gave his State of the Union address. Reagan's main topic was to be education, and he was expected to mention the shuttle and the first teacher in space, Christa McAuliffe.

"The shuttle solid rocket boosters (or SRBs) are key elements in the operation of the shuttle. Without the boosters, the shuttle cannot produce enough thrust to overcome the earth's gravitational pull and achieve orbit. There is an SRB attached to each side of the external fuel tank. Each booster is 149 feet long and 12 feet in diameter. Before ignition, each booster weighs 2 million pounds. Solid rockets in general produce much more thrust per pound than their liquid fuel counterparts. The drawback is that once the solid rocket fuel has been ignited, it cannot be turned off or even controlled. So it was extremely important that the shuttle SRBs were properly designed. Morton Thiokol [Inc] was awarded the contract to design and build the SRBs in 1974....

"The booster is comprised of seven hollow metal cylinders....The joints where the segments are joined together at KSC [Kennedy Space Center] are known as field joints....These field joints consist of a tang and clevis joint. The tang and clevis are held together by 177 clevis pins. Each joint is sealed by two O-rings....The second ring was added as a measure of redundancy since the boosters would be lifting humans into orbit. The purpose of the O-rings is to prevent hot combustion gases from escaping from the inside of the motor...."

The SRBs provide eighty percent of the thrust necessary to propel the shuttle into orbit and, about two minutes after launch, are designed to detach and parachute back to earth for use on future shuttle missions.

Figure 12-1 shows the ill-fated Challenger on the launch pad on the morning of January 28, 1986.

What Happened?

Within seconds after the launch of Challenger, puffs of smoke could be seen indicating that the O-rings on the right SRB were failing. In less than a minute a flame was evident (Figure 12-2). The flame struck the surface of the external tank that contained the liquid hydrogen and oxygen fuel for the shuttle's engines. The tank was breached and, 73 seconds after launch, Challenger exploded, claiming the lives of the seven members of the crew.

Figure 12-1. Challenger on the launch pad showing external tank (large tank behind the shuttle) and the SRBs (smaller cylinders on either side of the external tank). (Source: National Aeronautics and Space Administration.)

Figure 12-3 dramatically shows the result of the O-ring failure.

Figure 12-4 depicts the SRB joint that failed. The left illustration shows the hot gases (arrows) being shielded from the joint with zinc chromate putty. The right illustration shows what happened. The pressure caused a blowout in the putty, the gases penetrated through to the O-rings, eroding them and causing their failure.

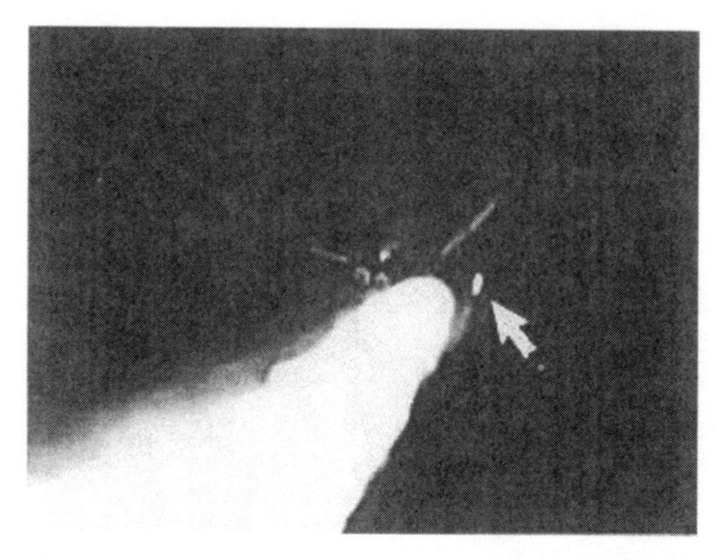

Figure 12-2. Flame impinging on the external fuel tank of Challenger. (Source: National Aeronautics and Space Administration.)

Figure 12-3. Explosion of Challenger. (Source: National Aeronautics and Space Administration.)

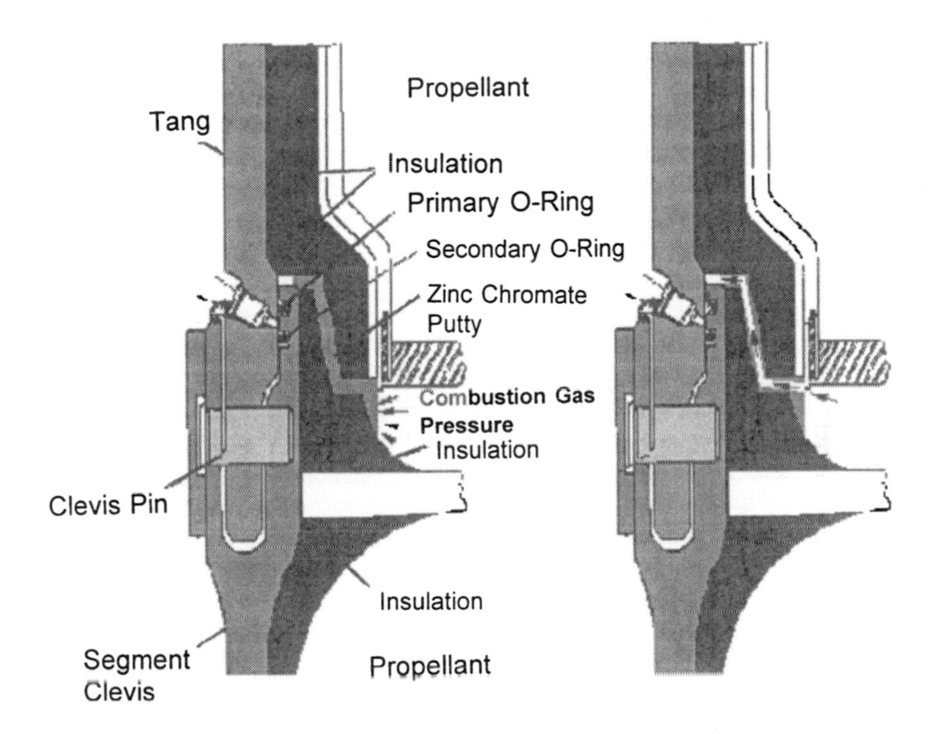

Figure 12-4. Solid rocket booster joint. (Source: Ref. 4, adapted from the Report of the Presidential Commission on the Space Shuttle Accident, p. 57.)

Figure 12-5 depicts the field joint, which failed on Challenger's right solid rocket booster.

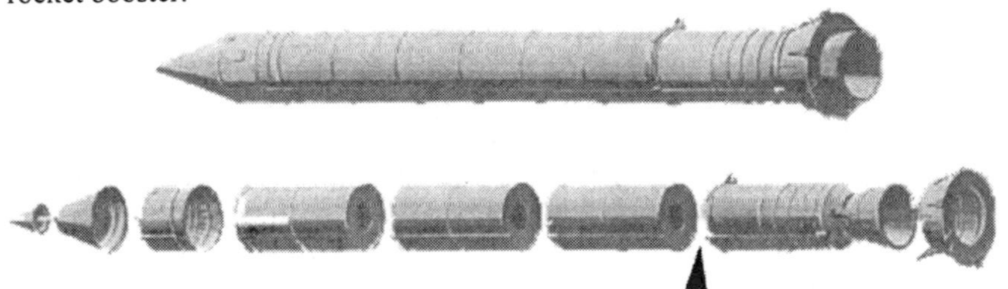

Figure 12-5. Solid rocket booster assembly. Arrow shows Challenger point of failure. (Source: Ref. 4, adapted from the Report of the Presidential Commission on the Space Shuttle Accident, p. 52.)

Why Did the O-rings Fail?

The Challenger mission was delayed because of a weather front expected to move into the area, bringing rain and cold temperatures. Weather forecasts indicated that record-setting low temperatures were probable. TAMU (3) reports what happened next as follows:

"NASA wanted to check with all of its contractors to determine if there would be any problems with launching in the cold temperatures. Alan McDonald, director of the Solid Rocket Motor Project at Morton Thiokol, was convinced that there were cold weather problems with the solid rocket motors and contacted two of the engineers working on the project, Robert Ebeling and Roger Boisjoly. Thiokol knew there was a problem with the boosters as early as 1977 and had initiated a redesign effort in 1985. NASA Level I management had been briefed on the problem on August 19, 1985. Almost half of the shuttle flights had experienced O-ring erosion in the booster field joints. Ebeling and Boisjoly had complained to Thiokol that management was not supporting the redesign task force.

"The size of the gap is controlled by several factors, including the dimensional tolerances of the metal cylinders and their corresponding tang or clevis, the ambient temperature, the diameter of the O-ring, the thickness of the shims, the loads on the segment, and quality control during assembly. When the booster is ignited, the putty is displaced, compressing the air between the putty and the primary O-ring....The air pressure forces the O-ring into the gap between the tang and clevis. Pressure loads are also applied to the walls of the cylinder, causing the cylinder to balloon slightly....This ballooning of the cylinder walls caused the gap between the tang and clevis gap to open. This effect has come to be known as joint rotation. Morton Thiokol discovered this joint rotation as part of its testing program in 1977. Thiokol discussed the problem with NASA and started analyzing and testing to determine how to increase the O-ring compression, thereby decreasing the effect of joint rotation. Three design changes were implemented:

1. Dimensional tolerances of the metal joint were tightened.
2. The O-ring diameter was increased, and its dimensional tolerances were tightened.
3. The use of the shims mentioned above was introduced. Further testing by Thiokol revealed that the second seal, in some cases, might not seal at all. Additional changes in the shim thickness and O-ring diameter were made to correct the problem.

"A new problem was discovered during November 1981, after the flight of the second shuttle mission. Examination of the booster field joints revealed that the O-rings were eroding during flight. The joints were still sealing effectively, but the O-ring material was being eaten away by hot gases that escaped past the putty. Thiokol studied different types of putty and its application to study their effects on reducing O-ring erosion. Shuttle flight 51-C of January 24, 1985, had been launched during some of the coldest weather in Florida history. Upon examination of the booster joints, engineers at Thiokol noticed black soot and grease on the outside of the booster casing, caused by actual gas blow-by. This prompted Thiokol to study the effects of O-ring resiliency at low temperatures....In July

1985, Morton Thiokol ordered new steel billets which would be used for a redesigned case field joint. At the time of the accident, these new billets were not ready for Thiokol, because they take many months to manufacture.

"Temperatures for the...launch date were predicted to be in the low 20°s. This prompted Alan McDonald to ask his engineers at Thiokol to prepare a presentation on the effects of cold temperature on booster performance. A teleconference was scheduled the evening before the rescheduled launch in order to discuss the low temperature performance of the boosters. This teleconference was held between engineers and management from Kennedy Space Center, Marshall Space Flight Center in Alabama, and Morton Thiokol in Utah. Boisjoly and another engineer, Arnie Thompson, knew this would be another opportunity to express their concerns about the boosters, but they had only a short time to prepare their data for the presentation. Thiokol's engineers gave an hour-long presentation, presenting a convincing argument that the cold weather would exaggerate the problems of joint rotation and delayed O-ring seating. The lowest temperature experienced by the O-rings in any previous mission was 53°F, the January 24, 1985 flight. With a predicted ambient temperature of 26°F at launch, the O-rings were estimated to be at 29°F. After the technical presentation, Thiokol's Engineering Vice President Bob Lund presented the conclusions and recommendations. His main conclusion was that 53°F was the only low temperature data Thiokol had for the effects of cold on the operational boosters. The boosters had experienced O-ring erosion at this temperature. Since his engineers had no low temperature data below 53°F, they could not prove that it was unsafe to launch at lower temperatures. He read his recommendations and commented that the predicted temperatures for the morning's launch...[were] outside the data base and NASA should delay the launch, so the ambient temperature could rise until the O-ring temperature was at least 53°F...Marshall's Solid Rocket Booster Project Manager...commented that the data was inconclusive and challenged the engineers' logic. A heated debate went on for several minutes before...Joe Kilminster [was asked] for his opinion. Kilminster was in management, although he had an extensive engineering background. By bypassing the engineers, ...[the project manager] was calling for a middle-management decision, but Kilminster stood by his engineers. Several other managers at Marshall expressed their doubts about the recommendations, and finally Kilminster asked for a meeting off of the net, so Thiokol could review its data. Boisjoly and Thompson tried to convince their senior managers to stay with their original decision not to launch. A senior executive at Thiokol...commented that a management decision was required. The managers seemed to believe the O-rings could be eroded up to one third of their diameter and still seat properly, regardless of the temperature. The data presented to them showed no correlation between temperature and the blow-by gases which eroded the O-rings in previous missions. According to testimony by Kilminster and Boisjoly, ...[the senior executive] finally turned to Bob Lund and said, 'Take off your engineering hat and put on your management hat.' Joe Kilminster wrote out the new recommendation and went back on line with the teleconference. The new recommendation stated that the cold was still a safety concern, but their people had found that the original data was indeed inconclusive and their 'engineering assessment' was that launch was recommended, even though the engineers had no part in writing the new recommendation and refused to sign it. Alan McDonald, who was present with NASA management in Florida, was surprised to see the recommendation to launch and appealed to NASA management not to launch. NASA managers decided to approve the boosters for launch despite the fact that the predicted launch temperature was outside of their operational specifications."

The rest is history.

Boisjoly Blows the Whistle

Roger Boisjoly, one of the engineers who had argued so strongly against the launch, had endeavored to do all in his power to work within the system to avert the tragedy but was overruled by a management decision. The damage was done and Boisjoly followed the *Code of Ethics* by blowing the whistle when he later testified before the Presidential Commission on the Space Shuttle Accident and told what had happened.

One year after the accident he related his experiences to faculty and students at Massachusetts Institute of Technology (4). These are his comments on the tragedy:

> It was approximately five minutes prior to launch as I was walking past the room used to view launches when Bob Ebeling stepped out to encourage me to enter and watch the launch. At first I refused, but he finally persuaded me to watch the launch. The room was filled, so I seated myself on the floor closest to the screen and leaned against Bob's legs as he was seated in a chair. The boosters ignited, and as the vehicle cleared the tower I whispered that we had just dodged a bullet. At approximately T+60 seconds Bob told me that he had just completed a prayer of thanks to the Lord for a successful launch. Just 13 seconds later we both saw the horror of destruction as the vehicle exploded. We all sat in stunned silence for a short time then I got up and left the room and went directly to my office where I remained the rest of the day. Two of my seal task-team colleagues inquired at my office to see if I was okay, but I was unable to speak to them and hold back my emotions so I just nodded yes to them and they left after a short silent stay.

Boisjoly consistently made ethical choices during his work for Morton Thiokol and NASA. He subsequently revealed what he knew to the Presidential Commission, despite opposing pressure from officials of his company. Weil (2) said, "If it is a distinguishing mark of actions labeled whistleblowing that the agent intends to force attention to a serious moral problem, both Boisjoly's and Mac-

Donalds's responses qualify. This feature is the foundation of the public's interest in whistleblowing. By bringing such serious problems to light, whistleblowers contribute to protecting the public's welfare. There have been instances of serious moral problems that were well known inside companies but did not get exposed for lack of a whistleblower....Organizations offer settings in which problems with potential for catastrophe can slowly ripen and somehow remain unattended to and unexposed even though many people in those settings are aware of the problem."

Roger Boisjoly survived this ordeal and made a new career. He subsequently sat for, and successfully passed, the professional engineering licensing examination. He later said, "I truly believe from my experience in the Challenger episode that if I had been licensed and threw that code of ethics in [management's] face, I think it would have got them to think" (6).

In 1988, in recognition of his efforts to avert the Challenger disaster, the American Association for the Advancement of Science presented Roger Boisjoly, PE, with the Award for Scientific Freedom and Responsibility.

Acknowledgement

The foregoing discussion of the Challenger disaster is based upon information obtained from the National Society of Professional Engineers (2), Texas A&M University (3), and the WWW Ethics Center for Engineering and Science and the National Science Foundation (4). More extensive information about the Challenger disaster may be found on the Internet at the addresses given in the list of references at the end of this chapter.

STUDY QUESTIONS

12.1 What could NASA management have done differently in the Challenger situation?

12.2 What, if anything, could their subordinates have done differently?

12.3 What should Roger Boisjoly have done differently (if anything)? In answering this question, keep in mind that, at his age, the prospect of finding a new job if he was fired was probably slim. He also had a family to support.

12.4 What do you see as your engineering professional responsibilities in relation to both being loyal to management and protecting the public welfare?

(The preceding four problems are from Ref. 3 with permission.)

REFERENCES

1. M Bright, J McCall, I Iyoob. The Whistleblowing Homepage. Oklahoma State University. Undated. <http://www.angelfire.com/ok/whistleblow>
2. V. Weil. Whistleblowing: What Have We Learned Since the Challenger? Center for the Study of Professional Ethics, Illinois Institute of Technology.

National Society of Professional Engineers. Undated.
<http://www.nspe.org/eh1-whi.htm>

3. The Space Shuttle Challenger Disaster. National Science Foundation Grant DIR-9012252, Departments of Philosophy and Mechanical Engineering, Texas A&M University. Undated.
<http://ethics.tamu.edu/ethics/shuttle/shuttle1.htm>
4. The Challenger Disaster. WWW Ethics Center for Engineering and Science. National Science Foundation. 4 October 1998.
<http://ethics.cwru.edu/boisjoly/RB1-0.html>
5. The Challenger Disaster. Undated.
<http://www.geocities.com/sunsetstrip/underground/8165>
6. E Kane. Is PE License a Boon to Ethics in Industry? Engineering Times 19 (3):1, March 1997.

13

International Ethics

A very pressing problem facing the engineering profession is how to work in a growing international business arena without compromising the codes of ethics of the profession. Engineering firms are increasingly engaged in multinational projects and face many situations which are unlike what they are accustomed to. Conditions are often different technically. Those can be readily handled. Ethical considerations, however, present even greater problems. Some of the ethical problems include:

- Inadequate environmental standards in comparison to the United States
- Sex discrimination
- Bribery to obtain favor with local officials
- Labor exploitation, in terms of wages, benefits, and safe working conditions
- Expected and expensive "gifts" to potential clients
- "Information brokering," i.e., selling of inside information to suppliers anxious to win contracts

I often give talks to various engineering groups about ethics and have conducted short courses on the subject. In some of these presentations I have presented the hypothetical situation of a multinational firm doing engineering work in a country where it is generally accepted business practice to give substantial gifts to government officials in order to obtain contracts, permits, or other actions without which a project could not proceed.

This is a very common practice in some countries and is totally legal under the laws of these countries. In fact the "gifts" are expected and without them the engineering firm often is not permitted to do the work. The "gifts" are pure and simply bribes, legal bribes in that country, but bribes nonetheless.

This situation presents an ethical dilemma for U.S. firms seeking foreign business because the practice is illegal in the United States and is clearly a violation of the NSPE *Code of Ethics for Engineers.* Section 5(b) of the *Code* reads, "Engineers shall not offer, give, solicit or receive, either directly or indirectly, any contribution to influence the award of a contract by public authority....They shall not offer any gift, or other valuable consideration in order to secure work. They shall not pay a commission, percentage or brokerage fee in order to secure work, except to bona fide employee or bona fide established commercial marketing agencies retained by them."

The dilemma here is that what is unacceptable and illegal in the U.S. is totally proper and legal in some other countries. What then should the engineering firm seeking foreign business do? I posed this problem to various groups and asked their opinion.

Almost without exception, they agreed that the practice of giving such gifts *by the engineer or engineering firm* is unethical, local custom and law not withstanding, and engineers should refrain from doing so even if it meant that the business would be lost to a competitor. However, at one presentation an engineer said his firm does a lot of business abroad and has solved the ethical dilemma by subcontracting with a local firm in the country in question and by letting the local firm handle gift giving. He reasoned that in this way he and his firm were not a party to the practice and were not acting unethically. The local firm which actually gave the gifts was similarly acting in an ethical manner since it was doing what was acceptable in its own country. Subsequently I have heard the same justification from other persons.

Is this a way out of the dilemma? This appears to be exactly what some U.S. firms do on a routine basis.

If you think this is a solution, look at Section 5(b) of the NSPE *Code* again. It begins, "Engineers shall not offer, give, solicit or receive, **either directly or indirectly** ...".

Clearly the use of a subcontractor to "do the dirty work" is indirectly giving the gift.

The practice is not ethical.

In Case No. 76-6 involving the practice of giving gifts to foreign officials, the NSPE Board of Ethical Review stated in part, "Even though the practice may be legal and accepted in the foreign country, and even though some might argue on pragmatic grounds the United States commercial companies should 'go along' to protect the jobs of employees in this country, we cannot accept it for professional services. No amount of rationalization or explanation will change the public reaction that the profession's claim of placing service before profit has been compromised by a practice which is repugnant to the basic principles of ethical behavior under the laws and customs of this country.... We believe that the code must be read on this most basic point of honor and integrity not only literally, but in the spirit of its purpose—to uphold the highest standards of the profession. Anything less is a rationalization which cannot stand the test of placing the public ahead of all other considerations."

If you work for a company which uses an agent to give gifts abroad, or if you are aware of a company which does this, you would do well to bring this discussion to their attention.

Engineering Times carried an editorial on this subject which bears repeating (1). It said in part:

> Engineers involved in an international business situation that strains their professional code of ethics may wish that an interpreter of ethics abroad were at least as accessible as a translator of languages.
>
> In the current absence of a uniform international approach, a good sense of ethics may feel like a handicap to professionals trying to succeed within cultures that play by different rules. If engineers "do as the Romans do" in violation of their ethical code, they may make shortsighted economic gains, but in the long run, they can lose the respect of their peers and the international community—and break the law.
>
> Bribing public officials to influence the award of projects is just one example of an unethical practice encountered overseas. Outlawed by the Federal Corrupt Practices Act in the U. S., such situations nevertheless arise in numerous countries. Development of an international code of ethics that could be accepted by the world community would be an unequivocal step toward preventing such policies.

Unfortunately, there is no international code of ethics which might help curtail those practices abroad which U.S. engineers see as being unethical. There is also no justification for "doing as the Romans do." Engineers must obey their codes of ethics, no matter what custom prevails at the project location. That may mean that the job will be lost in some cases but that is part of the price of practicing ethically.

Engineers who do think it is acceptable to use an agent abroad to "do the dirty work," thus acting like Pontius Pilate and "washing one's hands" of the problem, also need to realize that they are committing a felony under U.S. law and are subject to prosecution. At one seminar I conducted, a participant said the those who wish to engage in this practice have only one option, to give up their citizenship and become a citizen of the country in question. So long as they are U.S. citizens, the Federal Corrupt Practices Act applies to them. His comment also applies to corporations. A U.S. corporation is a citizen in the eyes of the law. Those companies engaging in bribery can avoid the law only by giving up their corporate charters and reincorporating abroad.

STUDY QUESTIONS

13.1 The following situation was presented in *Engineering Times* (2) in conjunction with the 1997 NSPE Chapter Ethics Program:

With the heightened globalization of engineering practice, professional engineers are increasingly finding themselves practicing in various countries and cultures and facing different standards of engineering ethics and practice. For example, in some countries, it is considered proper for government contractors and consultants to give expensive gifts to government officials before and after award of a contract. Building codes, environmental, health, and safety standards, and employment conditions may not be defined as rigorously as they are in the U.S. Engineering licensure laws and disciplinary procedures are also quite different. For those seeking to practice abroad, it is not always clear which standard should apply.

As a practicing PE abroad, what factors would you consider in deciding how to respond to these ethical challenges? What other ethical challenges do you think could arise from such differences in cultures? What standards should apply to you and your company when practicing within another nation? What would be the most appropriate way of addressing these and other issues?

13.2 Consider the following case from *Engineering Times* (3):

Avery Global, PE, is a consulting engineer who works in the U.S. and abroad. Global is contacted by the government of Country A and asked to submit a proposal for a major water project being constructed there. As a part of the project, Global is encouraged to associate with and retain Engineer B, a local engineer in Country A, with whom Global has worked on private projects in that country. One of the accepted "customs" in Country A is for consultants, such as engineers, to give substantial gifts to public officials in connection with awarding of public works contracts. Global recognizes that the giving of such gifts may be a violation of U.S. law—although it may not technically violate the law in Country A. Engineer B proposes to Global that if the project is awarded to Global's firm, Engineer B will handle "business arrangements" in Country A and that Global will be involved in overall project management as well as all technical matters.

Would it be ethical for Global to proceed with the project under these circumstances? If not, what action do you suggest that Global should take?

REFERENCES

1. International Ethics. Engineering Times 20(5):4, June 1998.
2. You Be the Judge - Chapter Ethics Contest. Engineering Times 19(7):3, July 1997.
3. You Be the Judge. Engineering Times 19(9), 3, Sept. 1997.

14

The First Fundamental Canon and the Second Mile

THE ENGINEER'S PRIMARY ETHICAL RESPONSIBILITY

The first Fundamental Canon of the NSPE *Code of Ethics* states:

> *Engineers, in the fulfillment of their professional duties shall hold paramount the safety, health and welfare of the public.*

The Canon overrides all other considerations. It does not say "may." It says, "shall." It does not leave any options open. The engineer cannot rationalize or justify compromises to public health, welfare, and safety. No matter what the business or economic considerations may be, the safety, health, and welfare of the public is of paramount importance.

This emphasis is not unique to the NSPE *Code of Ethics*. Consider what the *Code of Ethics* of the Institute of Electrical and Electronics Engineers says. Its opening Canon requires IEEE members

> *to accept responsibility in making engineering decisions consistent with the safety, health and welfare of the public, and to disclose promptly disclose factors that might endanger the public or the environment.*

The Canon not only requires the electrical engineer to give first responsibility to public welfare but also requires the engineer to "whistleblow" when others do not do likewise.

The American Society of Civil Engineers *Code of Ethics* also enumerates a series of Fundamental Canons, the first of which reads:

Engineers shall hold paramount the safety, health and welfare of the public and shall strive to comply with the principles of sustainable development in the performance of their professional duties.

This ASCE Canon and the IEEE Canon go beyond the NSPE Canon by specifically placing an obligation upon engineers to shown due concern for the environmental consequences of their actions.

AACE International, the Association for the Advancement of Cost Engineering, echoes this with the words:

Members will hold paramount the safety, health, and welfare of the public, including that of future generations.

The codes of ethics of other engineering societies place comparable emphasis on public, health, and safety and many of them stress long term effects of an engineers actions as do the IEEE, ASCE, and AACE codes.

The same emphasis also commonly appears in codes of ethics of other organizations that are not exclusively engineering oriented. The Project Management Institute is a very multidisciplinary organization that is composed of members from a great many professions, not engineering alone. Their *Code of Ethics*, Article IV reads:

Project management professionals shall, in fulfilling their responsibilities to the community, protect the safety, health, and welfare of the public and speak out against abuses in those areas affecting the public interest.

All of these various codes of ethics appear in the Appendix of this book. Read through them and the recurrent theme of public health, safety, and welfare, as the primary obligation of professionals is evident throughout the document—not only in the sentences quoted above.

The *Engineers Creed* which appears in Chapter 6 states that professional engineers dedicate their skills to the "advancement and betterment of human welfare" and pledge "to place service before profit, the honor and standing of the profession before personal advantage, and the public welfare above all other considerations".

Unfortunately the *Engineers Creed* and the codes of ethics do not always prevail as has been illustrated several times in previous chapters of this book, and as will be further illustrated in Chapter 15, Selected Case Studies, the final chapter of this book.

THE HYATT REGENCY WALKWAYS COLLAPSE: ETHICAL ISSUES OF THE CASE*

On July 17, 1981, the Hyatt Regency Hotel in Kansas City, Missouri held a videotaped tea-dance party in their atrium lobby. With many partygoers standing and dancing on the suspended walkways, connections supporting the ceiling rods that held up the second and fourth-floor walkways across the atrium failed, and both walkways collapsed onto the crowded first-floor atrium below. The fourth-floor walkway collapsed onto the second-floor walkway, while the offset third-floor walkway remained intact. As the United States' most devastating structural failure, in terms of loss of life and injuries, the Kansas City Hyatt Regency walkways collapse left 114 dead and in excess of 200 injured. In addition, millions of dollars in costs resulted from the collapse, and thousands of lives were adversely affected.

The hotel had only been in operation for approximately one year at the time of the walkways collapse, and the ensuing investigation of the accident revealed some unsettling facts (2).

First, during January and February 1979, over a year before the collapse, the design of the walkway hanger rod connections was changed in a series of events and communications (or disputed miscommunications) between the fabricator and the engineering design team, a professional engineering firm. The fabricator changed the design from a one-rod to a two-rod system to simplify the assembly task, doubling the load on the connector, which ultimately resulted in the walkways collapse.

Second, the fabricator, in sworn testimony before the administrative judicial hearings after the accident, claimed that his company telephoned the engineering firm for change approval. The engineering firm denied ever receiving such a call.

Third, on October 14, 1979, while the hotel was still under construction, more than 2700 square feet of the atrium roof collapsed because one of the roof connections at the north end of the atrium failed. In testimony, the engineering firm stated that on three separate occasions they requested on-site project representation to check all fabrication during the construction phase; however, these requests were not acted on by the owner, a redevelopment company, due to additional costs of providing on-site inspection.

Fourth, even as originally designed, the walkways were barely capable of holding up the expected load, and would have failed to meet the requirements of the Kansas City Building Code.

The design engineering firm was responsible for preparing structural engineering drawings for the Hyatt project: three walkways spanning the atrium area of the hotel. Wide flange beams with 16-inch depths (W16 x 26) were used along

----

*Abridged by permission from a Texas A&M University engineering ethics case study prepared under National Science Foundation Grant No. DIR-90112252. See Ref. 1.

either side of the walkway and hung from a box beam (made from two MC 8 x 8.5 rectangular channels, welded toe-to-toe). A clip angle welded to the top of the box beam connected these beams by bolts to the W section. This joint carried virtually no moment, and therefore was modeled as a hinge. One end of the walkway was welded to a fixed plate and would be a fixed support, but for simplicity, it could be modeled as a hinge. This only makes a difference on the hanger rod nearest this support (it would carry less load than the others do and would not govern design). The other end of the walkway support was a sliding bearing modeled by a roller. The original design for the hanger rod connection to the fourth-floor walkway was a continuous rod through both walkway box beams.

Events and disputed communications between the engineering firm and the fabricators resulted in a design change from a single to a double hanger rod box beam connection for use at the fourth-floor walkways. The fabricator requested this change to avoid threading the entire rod. They made the change, and the contract's Shop Drawing 30 and Erection Drawing E-3 were changed. The effect of the change was that the bolt under the top walkway had to support not only the top walkway but also the one hung from it. This effectively doubled the load on that bolt.

On February 16, 1979, the engineering firm received 42 shop drawings and returned them to the fabricator 10 days later, stamped with a professional engineer's seal, authorizing construction. The fabricator built the walkways in compliance with the directions contained in the structural drawings, as interpreted by the shop drawings, with regard to these hangers. In addition, the fabricator followed the American Institute of Steel Construction (AISC) guidelines and standards for the actual design of steel-to-steel connections by steel fabricators.

As a precedent for the Hyatt case, the *Guide to Investigation of Structural Failure*, states that:

"Overall collapses resulting from connection failures have occurred only in structures with few or no redundancies. Where low-strength connections have been repeated, the failure of one has lead to failure of neighboring connections and a progressive collapse has occurred. The primary causes of connection failures are:

1. Improper design due to lack of consideration of all forces acting on a connection, especially those associated with volume changes.
2. Improper design utilizing abrupt section changes resulting in stress concentrations.
3. Insufficient provisions for rotation and movement.
4. Improper preparation of mating surfaces and installation of connections.
5. Degradation of materials in a connection.
6. Lack of consideration of large residual stresses resulting from manufacture or fabrication." (3)

On October 14, 1979, part of the atrium roof collapsed while the hotel was under construction. As a result, the owner called in the inspection team. The inspection team's contract dealt primarily with the investigation of the cause of the roof collapse and created no obligation to check any engineering or design work beyond the scope of their investigation and contract. In addition to the inspection team, the owner retained, on October 16, 1979, an independent engineering firm, to investigate the cause of the atrium roof collapse. On October 20, 1979, the engineering firm which had designed the project wrote the owner, stating that it was undertaking both an atrium collapse investigation as well as a thorough design check of all the members comprising the atrium roof. The engineering firm promised to check **all** steel connections in the structures, not just those found in the roof.

From October to November 1979, various reports were sent from the engineering firm to the owner and architect, assuring the overall safety of the entire atrium. In addition to the reports, meetings were held between the owner, architect and the engineers.

In July of 1980, the construction was complete, and the Kansas City Hyatt Regency Hotel was open for business.

Just one year later, on July 17, 1981, the box beams resting on the supporting rod nuts and washers were deformed, so that the box beam resting on the nuts and washers on the rods could no longer hold up the load. The box beams (and walkways) separated from the ceiling rods and the fourth- and second-floor walkways across the atrium of the Hyatt Regency Hotel collapsed

The following sequence of photographs documents the aftermath of the collapse. These photographs were taken shortly after the collapse by Dr. Lee Lowery, Jr., PE, of the Texas A&M University.

Figure 14-1. Photo of walkway sections taken from second-floor opening. Walkway sections have been moved from their original positions during the attempt to extricate those trapped in the wreckage.

Figure 14-2. Photo of still hanging third-floor walkway. Note the freestanding stairs from the second to the third floor in the background. With its columnless design, the stairs seem to be floating in air. The lobby was indeed a masterpiece of architecture and engineering, which had it been executed properly, would have provided its owners with profit and the public with a stunning atmosphere for years.

Figure 14-3. General view of the lobby floor during the first day of the investigation.

Figure 14-4. Close-up photo of the hanger rod threads, washer, and supporting nut. Note the deformation caused in the washer as the beam slipped around it.

Figure 14-5. Photo of one of the walkway crossbeams, lying on the floor of the lobby. This is one of the fourth-floor beams, as evidenced by having two bolt holes drilled through the beam. The second-floor beams had a single rod hole.

Figure 14-6. Close-up of one of the fourth-floor beams.

Following the accident investigations, on February 3, 1984, the Missouri Board of Architects, Professional Engineers and Land Surveyors filed a complaint against the design engineers, charging gross negligence, incompetence, misconduct, and unprofessional conduct in the practice of engineering in connection with their performance of engineering services in the design and construction of the Hyatt Regency Hotel. The report noted:

The hanger rod detail actually used in the construction of the second and fourth-floor walkways is a departure from the detail shown on the contract drawings. In the original arrangement each hanger rod was to be continuous from the second-floor walkway to the hanger rod bracket attached to the atrium roof framing. The design load to be transferred to each hanger rod at the second-floor walkway would have been 20.3 kips (90 kN). An essentially identical load would have been transferred to each hanger rod at the fourth-floor walkway. Thus the design load acting on the upper portion of a continuous hanger rod would have been twice that acting on the lower portion, but the required design load for the box beam hanger rod connections would have been the same for both walkways (20.3 kips (90 kN)).

The hanger rod configuration actually used consisted of two hanger rods: the fourth-floor-to-ceiling hanger rod segment as originally detailed on the second-to fourth-floor segment which was offset 4 in. (102 mm) inward along the axis of the box beam. With this modification the design load to be transferred by each second-floor box beam-hanger rod connection was unchanged, as were the loads in the upper and lower hanger rod segments. However, the load to be transferred from the fourth-floor box beam to the upper hanger rod under this arrangement was essentially doubled, thus compounding an already critical condition. The design load for a fourth-floor box beam-hanger rod connection would be 40.7 kips (181 kN) for this configuration.

Had this change in hanger rod detail not been made, the ultimate capacity of the box beam-hanger rod connection still would have been far short of that expected of a connection designed in accordance with the Kansas City Building Code, which is based on the AISC Specification. In terms of ultimate load capacity of the connection, the minimum value should have been 1.67 times 20.3, or 33.9 kips (151 kN). Based on test results the mean ultimate capacity of a single-rod connection is approximately 20.5 kips (91 kN), depending on the weld area. Thus the ultimate capacity actually available using the original connection detail would have been approximately 60% of that expected of a connection designed in accordance with AISC Specifications. Figures 14-7 and 14-8 illustrate potential failure of the third-floor walkway that did not collapse.

Figure 14-7. Photo of third-floor walkway connections from below. See Figure 14-2 for an overall view of the third-floor walkway. Note that from a distance, the fact that the third-floor walkway was also distressed was not apparent. Also, the fireproofing cover box has been removed at this time.

During the 26-week administrative law trial that ensued, representatives of the engineering firm denied ever receiving the call about the design change. Yet, one of their engineers affixed his seal of approval to the revised engineering design drawings.

Results of the hearing concluded that the engineering firm, in preparation of their structural detail drawings, "depicting the box beam hanger rod connection for the Hyatt atrium walkways, failed to conform to acceptable engineering practice. [This is based] upon evidence of a number of mistakes, errors, omissions and inadequacies contained on this section detail itself and of [the engineering firm's] alleged failure to conform to the accepted custom and practice of engineering for proper communication of the engineer's design intent."

Evidence showed that neither due care during the design phase, nor appropriate investigations following the atrium roof collapse were undertaken by the engineering firm. In addition, the firm was found responsible for the change from a one-rod to a two-rod system. Further, it was found that even if the fabricator failed to review the shop drawings or to specifically note the box beam hanger rod connections, the engineers were still responsible for the final check. Evidence showed that the engineers did not "spot check" the connection or the atrium roof collapse, and that they placed too much reliance on the fabricator.

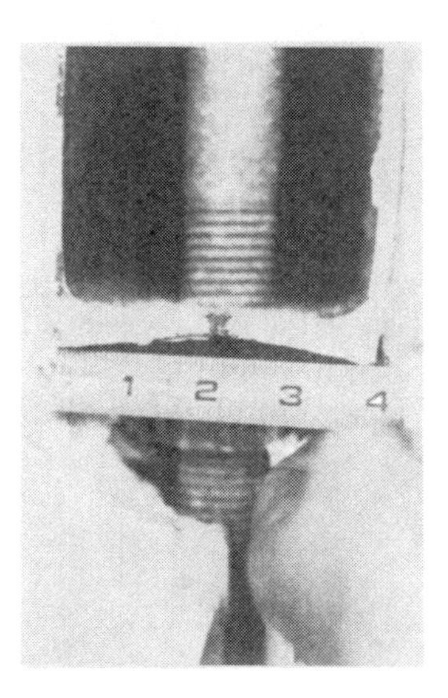

Figure 14-8. Close-up of third-floor hanger rod and crossbeam, showing yielding of the material. The flanges have been bent significantly, and the webs are bowed out against the fireproofing sheet rock. It should be remembered that the third-floor-walkway crossbeams were subjected to only half the loading of that induced in the fourth-floor beams. The distortion shown below was caused by only very light loading, mostly due to the dead load of the structure.

Due to evidence supplied at the hearings, a number of principals involved lost their engineering licenses, a number of firms went bankrupt, and many expensive legal suits were settled out of court. In November 1984, two engineers and the engineering firm were found guilty of gross negligence, misconduct, and unprofessional conduct in the practice of engineering. Subsequently, the engineers lost their licenses to practice engineering in the State of Missouri (and later, Texas), and the engineering firm had its certificate of authority as an engineering firm revoked.

As a result of the Hyatt Regency Walkways Collapse, the American Society of Civil Engineers (ASCE) adopted a report that states structural engineers have full responsibility for design projects.

Both of the engineers who lost their licenses are now practicing engineers in states other than Missouri and Texas.

THE SECOND MILE

The NSPE *Opinions of the Board of Ethical Review* (4) states that "Ethics provides the framework within which engineers may travel 'the second mile.'" And quotes some comments made by Dr. William E. Wilhenden of Case Institute of Technology. Dr. Wilhenden's comments are a fitting conclusion for this chapter. He said:

> *Every calling has its mile of compulsion: its round of tasks and duties, its prescribed man-to-man relationships, which one must traverse daily if one is to survive.*
>
> *Beyond that is the mile of voluntary effort where one strives for special excellence, seeks self-expression more than material gain, and gives that unrequited margin of service to the common good which invests work with a wide and enduring significance.*
>
> *The best fun of life and most of its durable satisfaction lies in the second mile and it is only here that a calling can attain the dignity and distinction of a professional.*

Whether you are an engineering student, a graduate engineer, or someone else with an interest in engineering ethics, it is hoped that this book has proven to be beneficial to you and has helped to broaden your understanding of engineering ethics in general and specifically of the first Fundamental Canon.

A COMPREHENSIVE STUDY QUESTION

14.1 The Kansas City Hyatt Regency walkway case centers on the question of who is responsible for design failure. As an ethical issue:

Who is ultimately responsible for checking the safety of final designs as depicted in shop drawings?

When we take the implicit social contract between engineers and society, the issue of public risk and informed consent, and codes of ethics of professional societies into account, it seems clear that the engineer must assume this responsibility when any change in design involving public safety carries a licensed engineer's signature. Yet, in terms of meeting building codes, what are the responsibilities of the engineer? The fabricator? The owner?

If we assume the engineer in the Hyatt case received the fabricator's telephone call requesting a verbal approval of the design change for simplifying assembly, what would make him approve such an untenable change? Some possible reasons include:

- saving time;
- saving money;
- avoiding a call for reanalysis, thereby raising the issue of a request to recheck all connector designs following the previous year's atrium roof collapse;
- following his immediate supervisor's orders;
- looking good professionally by simplifying the design;
- misunderstanding the consequences of his actions; or
- any combination of the above.

These reasons do not, however, fall within acceptable standards of engineering professional conduct. Instead, they pave the way for legitimate charges of negligence, incompetence, misconduct, and unprofessional conduct in the practice of engineering. When the engineer's actions are compared to professional responsibilities cited in the engineering codes of ethics, an abrogation of professional responsibilities by the engineer in charge is clearly demonstrated. But what of the owner, or the fabricator?

What if the call was not made? While responsibility rests with the fabricator for violating building codes, would the engineers involved in the case be off the hook? Why or why not?

The Hyatt Regency walkways collapse has resulted in a nationwide reexamination of building codes. In addition, professional codes on structural construction management practices are changing in significant ways.

Finally, what is your assessment of this case, based on the following questions:

What measures can professional societies take to ensure catastrophes like the Hyatt Regency Walkways Collapse do not occur?

Should the engineers allowed to practice engineering in other states? Why or why not? What is the engineering society's responsibility in this realm?

REFERENCES

1. The Kansas City Hyatt Regency Walkways Collapse, National Science Foundation Grant DIR-901225, Departments of Philosophy and Mechanical Engineering, Texas A&M University, Undated. <http://ethics.tamu.edu/ethics/hyatt/hyatt1.htm>
2. Missouri Board for Architects, Professional Engineers and Land Surveyors vs. Daniel M. Duncan, Jack D. Gillum and G.C.E. International, Inc., before the Administrative Hearing Commission, State of Missouri, Case No.

AR840239, Statement of the Case, Findings of Fact, Conclusions of Law and Decision rendered by Judge James B. Deutsch, November 14, 1985.
3. JR Janney (ed.). Guide to Investigation of Structural Failures. American Society of Civil Engineers' Research Council on Performance of Structures. Sponsored by the Federal Highway Administration, U.S. Department of Transportation, Contract No. DOTFH118843, 1979.
4. Opinions of the Board of Ethical Review, Vols. 2 - 7. Alexandria, VA: National Society of Professional Engineers, 1998, p.3 (each volume).

15

Selected Case Studies

INTRODUCTION

This concluding Chapter presents some case studies selected from among the various rulings of the NSPE Board of Ethical Review. The cases are broadly classified as dealing with:

- public safety and welfare;
- conflicts of interest;
- ethical trade practices;
- international ethics;
- research ethics.

The classification of some of the cases is somewhat arbitrary because they could fit into more than one of the above classifications.

Because the NSPE *Code of Ethics* has been revised several times over the years, to avoid confusion, the case studies have been modified as necessary to correspond with the July 1996 version of the *Code* which is included in the Appendix. In most cases, the changes were nothing more than changing reference numbers for various sections of the *Code*. However, the text has been changed slightly in instances where the Board referred to something which was in an earlier version of the *Code* but which does not appear in the current version. These various changes, however, do not modify in any way the conclusions and recommendations of the Board of Ethical Review (BER).

DISCLAIMER

All of the cases in this chapter have been furnished by the Board of Ethical Review of the National Society of Professional Engineers. The opinions are based upon data submitted to the BER and do not necessarily represent all of the pertinent facts when applied to any specific case. These opinions are for educational purposes only and should not be construed as expressing any opinion on the ethics of specific individuals.

In regard to the question of application of the NSPE *Code of Ethics* to corporations vis-à-vis real persons, business form or type should not negate nor influence conformance of individuals to the *Code*. The *Code* deals with professional services, which services must be performed by real persons. Real persons in turn establish and implement policies within business structures. The NSPE *Code* and the codes of other engineering societies are clearly written to apply to engineers, and it is incumbent on members of engineering societies to live up to the provisions of the codes of the societies of which they are members. This applies to all pertinent sections of the various codes.

PUBLIC HEALTH AND SAFETY CASE STUDIES

Hazardous Waste Disposal (BER Case 92-6)

Facts

Technician A is a field technician employed by a consulting environmental engineering firm. At the direction of his supervisor Engineer B, Technician A samples the contents of drums located on the property of a client. Based on Technician A's past experience, it is his opinion that analysis of the sample would most likely determine that the drum contents would be classified as hazardous waste. If the material is hazardous waste, Technician A knows that certain steps would legally have to be taken to transport and properly dispose of the drum including notifying the proper federal and state authorities.

Technician A asks his supervisor Engineer B what to do with the samples. Engineer B tells Technician A only to document the existence of the samples. Technician A is then told by Engineer B that since the client does other business with the firm, Engineer B will tell the client where the drums are located but do nothing else. Thereafter, Engineer B informs the client of the presence of drums containing "questionable material" and suggests that they be removed. The client contacts another firm and has the material removed.

Questions:

1. Was it ethical for Engineer B to merely inform the client of the presence of the drums and suggest that they be removed?
2. Did Engineer B have an ethical obligation to take further action?

Discussion

The extent to which an engineer has an obligation to hold paramount the public health and welfare in the performance of professional duties (Section I.1.) has been widely discussed by the Board of Ethical Review over the years. In many of these cases this basic duty has frequently intersected with the duty of engineers not to disclose confidential information concerning the business affairs, etc., of clients (Section III.4.).

For example, in BER Case 89-7 an engineer was retained to investigate the structural integrity of a 60-year old occupied apartment building which his client was planning to sell. Under the terms of the agreement with the client, the structural report written by the engineer was to remain confidential. In addition, the client made it clear to the engineer that the building was being sold "as is" and the client was not planning to take any remedial action to repair or renovate any system within the building. The engineer performed several structural tests on the building and determined that the building was structurally sound. However, during the course of providing services, the client confided in the engineer that the building contained deficiencies in the electrical and mechanical systems which violated applicable codes and standards. While the engineer was neither an electrical or mechanical engineer, he did realize that those deficiencies could cause injury to the occupants of the building and so informed the client. In his report, the engineer made a brief mention of his conversation with the client concerning the deficiencies; however, in view of the terms of the agreement, the engineer did not report the safety violations to any third parties. In determining that it was unethical for the engineer not to report the safety violations to appropriate public authorities, the Board, citing cases decided earlier, noted that the engineer "did not force the issue but instead went along without dissent or comment. If the engineer's ethical concerns were real, the engineer should have insisted that the client take appropriate action or refuse to continue work on the project." The Board concluded that the engineer had an obligation to go further, particularly because the *Code* uses the term "paramount" to describe the engineer's obligation to protect the public safety health and welfare.

More recently, in BER Case 90-5, the Board reaffirmed the basic principle articulated in BER Case 89-7. There, tenants of an apartment building sued its owner to force him to repair many of the building's defects. The owner's attorney hired an engineer to inspect the building and give expert testimony in support of the owner. The engineer discovered serious structural defects in the building which he believed constituted an immediate threat to the safety of the tenants. The tenants' suit had not mentioned these safety-related defects. Upon reporting the findings to the attorney, the engineer was told he must maintain this information as confidential as it is part of the lawsuit. The engineer complies with the request. In deciding it was unethical for the engineer to conceal his knowledge of the safety-related defect, the Board discounted the attorney's statement that the engineer was legally bound to maintain confidentiality, noting that any such duty was superseded by the immediate and imminent danger to the building's tenants. While the

Board recognized that there might be circumstances where the natural tension between the engineer's public welfare responsibility and the duty of nondisclosure may be resolved in a different manner, the Board concluded that this clearly was not the case under the facts.

Turning to the facts in this case, we believe the basic principles enunciated in BER Cases 89-7 and 90-5 are applicable here as well except in a different context. Unlike the facts in the earlier cases, Engineer B made no oral or written promise to maintain the client's confidentiality. Instead, Engineer B consciously and affirmatively took actions that could cause serious environmental danger to workers and the public, and also a violation of various environmental laws and regulations. Under the facts, it appears that Engineer B's primary concern was not so much maintaining the client's confidentiality as it was in maintaining good business relations with a client. In addition, it appears that as in all cases which involve potential violations of the law, Engineer B's actions may have had the effect of seriously damaging the long-term interests and reputation of the client. In this regard, we would also note that under the facts it appears that the manner in which Engineer B communicated the presence of the drums on the property must have suggested to the client that there was a high likelihood that the drums contained hazardous materials. We believe that this subterfuge is wholly inconsistent with the spirit and intent of the *Code of Ethics* because it makes the engineer an accomplice to what may amount to an unlawful action.

Clearly, Engineer B's responsibility under the facts was to bring the matter of the drums possibly containing hazardous material to the attention of the client with a recommendation that the material be analyzed. To do less would be unethical. If analysis demonstrates that the material is indeed hazardous, the client would have the obligation of disposing of the material in accordance with applicable federal state and local laws.

Conclusions

1. It was unethical for Engineer B to merely inform the client of the presence of the drums.
2. It was unethical for Engineer B to fail to advise his client that he suspected hazardous material and provide a recommendation concerning removal and disposal in accordance with federal, state, and local laws.

Responsibilities of a Government Engineer (BER Case 92-4)

Facts

Engineer A, an environmental engineer employed by the state environmental protection division, is ordered to draw up a construction permit for construction of a power plant at a manufacturing facility. He is told by a superior to move expeditiously on the permit and "avoid any hang-ups" with respect to technical issues. Engineer A believes the plans as drafted are inadequate to meet the regulation requirements and that outside scrubbers to reduce sulfur dioxide emissions are necessary and without them the issuance of the permit would violate certain air pollution standards as mandated under the 1990 Clear Air Act. His superior believes that plans which involve limestone mixed with coal in a fluidized boiler process would remove 90% of the sulfur dioxide will meet the regulatory requirements. Engineer A contacts the state engineering registration board and is informed, based upon the limited information provided to the board that suspension or revocation of his engineering license was a possibility if he prepared a permit that violated environmental regulations. Engineer A refused to issue the permit and submitted his findings to his superior. The department authorized the issuance of the permit. The case had received widespread publicity in the news media and is currently being investigated by state authorities.

Questions

1. Would it have been ethical for Engineer A to withdraw from further work in this case?
2. Would it have been ethical for Engineer A to issue the permit?
3. Was it ethical for Engineer A to refuse to issue the permit?

Discussion

The facts of this case are in many ways a classic ethical dilemma faced by many engineers in their professional lives. Engineers have a fundamental obligation to hold paramount the safety, health, and welfare of the public in the performance of their professional duties (*Code* Section I.1.). Moreover, the *Code* provides guidance to engineers who are confronted with circumstances where their professional reputation is at stake. Sometimes engineers are asked by employers or clients to sign off on documents in which they may have reservations or concerns.

The Board of Ethical Review has examined this issue over the years in differing contexts. As early as case BER 65-12, the Board dealt with a situation in which a group of engineers believed that a product was unsafe. The Board then determined that as long as the engineers held to that view, they were ethically justified in refusing to participate in the processing or production of the product in question. The Board recognized that such action by the engineers would likely lead to loss of employment.

In BER Case 82-5, where an engineer employed by a large defense industry firm documented and reported to his employer excessive costs and time delays by subcontractors, the Board ruled that the engineer did not have an ethical obligation to continue his efforts to secure a change in the policy after his employer rejected his reports, or to report his concerns to proper authority, but has an ethical right to do so as a matter of personal conscience. The Board noted that the case did not involve a danger to the public health or safety, but related to a claim of unsatisfactory plans and the unjustified expenditure of public funds. The Board indicated that it could dismiss the case on the narrow ground that the *Code* does not apply to a claim not involving public health and safety, but that was too narrow a reading of the ethical duties of engineers engaged in such activities. The Board also stated that if an engineer feels strongly that an employer's course of conduct is improper when related to public concerns, and if the engineer feels compelled to blow the whistle to expose facts as he sees them, he may well have to pay the price of loss of employment. In this type of situation, the Board felt that the ethical duty or right of the engineer becomes a matter of personal conscience, but the Board was unwilling to make a blanket statement that there is an ethical duty in these kinds of situations for the engineer to continue the campaign within the company, and make the issue one for public discussion.

More recently, in BER Case 88-6, an engineer was employed as the city engineer/director of public works with responsibility for disposal plants and beds and reported to a city administrator. After (l) noticing problems with overflow capacity which are required to be reported to the state water pollution control authorities, (2) discussing the problem privately with members of the city council, (3) being warned by the city administrator to only report the problem to him, (4) discussing the problem again informally with the city council, and (5) being relieved by the city administrator of responsibility for the disposal plants and beds by a technician, the engineer continued to work in the capacity as city engineer/director of public works. In ruling that the engineer failed to fulfill her ethical obligations by informing the city administrator and certain members of the city council of her concern, the Board found that the engineer was aware of a pattern of ongoing disregard for the law by her immediate supervisor as well as by members of the city council. After several attempts to modify the views of her superiors, the engineer knew or should have known that "proper authorities" were not the city officials, but more probably state officials. The Board could not find it credible that a city engineer/director of public works for a medium-sized town would not be aware of this basic obligation. Said the Board, the engineer's inaction permitted a serious violation of the law to continue and made the engineer an "accessory" to the actions of the city administrator and others.

Turning to the facts of this case, we believe the situation involved in this case is in many ways similar to the situation involved in BER Case 88-6. This, unlike BER Case 82-5 did not involve a matter of personal conscience, but rather a matter which had a direct impact upon the public health and safety. Yet unlike the circumstances involved in BER Case 88-6 where the issues were hidden from public note, here the case involves facts which have received coverage in the media. In view of this fact, we do not believe it is incumbent upon Engineer A to bring this issue to the attention of the "proper authorities." As we see it, such officials are already aware of the situation and have begun an investigation. The reason for our position in BER Case 88-6 was that the engineer's failure to bring the problems to the attention of the "proper authorities" made it more probable that danger would ultimately result to the public health, safety, and welfare. Here, appropriate public officials presumably already know the circumstances. To bring the matter to their attention is a useless act.

However, we believe it would not have been ethical for Engineer A to withdraw from further work on the project because Engineer A had an obligation to stand by his position consistent with his obligation to protect the public health, safety, and welfare and refuse to issue the permit. Engineers have an essential role as technically qualified professionals to "stick to their guns" and represent the public interest under the circumstances where they believe the public health and safety is at stake.

We would also note that this case also raises another dimension which involves the role of the state licensing board in determining the ethical conduct of licensees. Under the facts, Engineer A affirmatively sought the opinion of the state as to whether his approval of the permit could violate the state engineering registration law. We believe Engineer A's actions in this regard constitute appropriate conduct and actions are consistent with Section II.l.a. of the *Code*. This case involves a question of public health and welfare and Engineer A's decision to disassociate himself from further work on this project avoids having Engineer A being placed in a professionally compromising situation.

Conclusions

1. It would not have been ethical for Engineer A to withdraw from further work on the project.
2. It would not have been ethical for Engineer A to issue the permit.
3. It was ethical for Engineer A to refuse to issue the permit.

Whistleblowing City Engineer (BER Case 88-6)

Facts

Engineer A is employed as the City Engineer/Director of Public Works for a medium-sized city and is the only licensed professional engineer in a position of responsibility in the city government. The city has several large food processing plants that discharge very large amounts of vegetable wastes into the city's sanitary system during the canning season. Part of the canning season coincides with the rainy season.

Engineer A has the responsibility for the disposal plant and beds and is directly responsible to City Administrator C. Technician B answers to Engineer A.

During the course of her employment, Engineer A notifies Administrator C of the inadequate capacity of the plant and beds to handle the potential overflow during the rainy season and offers possible solutions. Engineer A has also discussed the problem privately with certain members of the city council without the permission of City Administrator C. City Administrator C has told Engineer A that "we will face the problem when it comes." City Administrator C orders Engineer A to discuss the problems only with him and warns her that her job is in danger if she disobeys.

Engineer A again privately brings the problem up to other city officials. City Administrator C removes Engineer A from responsibility of the entire sanitary system and the chain of command by a letter instructing Technician B that he is to take responsible charge of the sanitary system and report directly to City Administrator C. Technician B asks for a clarification and is again instructed via memo by City Administrator C that he, Technician B, is completely responsible and is to report any interference by a third party to City Administrator C. Engineer A receives a copy of the memo. In addition, Engineer A is placed on probation and ordered not to discuss this matter further and that if she does she will be terminated.

Engineer A continues in her capacity as City Engineer/Director of Public Works, assumes no responsibility for the disposal plant and beds, but continues to advise Technician B without the knowledge of City Administrator C.

That winter during the canning season, particularly heavy storms occur in the city. It becomes obvious to those involved that if wastewater from the ponds containing the domestic waste is not released to the local river, the ponds will overflow the levees and dump all waste into the river. Under state law, this condition is required to be reported to the state water pollution control authority, the agency responsible for monitoring and overseeing water quality in state streams and rivers.

Question

Did Engineer A fulfill her ethical obligation by informing City Administrator C and certain members of the city council of her concerns?

Discussion

The engineer's obligation to hold paramount the safety, health, and welfare of the public in the performance of his professional duties is probably among the most basic. Clearly, its importance is evident by the fact that it is the very first obligation stated in the NSPE *Code of Ethics*. Moreover, the premise upon which professional engineering exists—the engineering registration process—is founded upon the proposition that in order to protect the public health and safety, the state has an interest in regulating by law the practice of the profession.

While easily stated in the abstract, the breadth and scope of this fundamental obligation is far more difficult to fix. As we have long known, ethics frequently involves a delicate balance between competing and, oft times, conflicting obligations. However, it seems clear that where the conflict is between one important obligation or loyalty and the protection of the public, for the engineer the latter must be viewed as the higher obligation.

The Board has faced this most difficult issue on two other occasions in somewhat dissimilar circumstances. In Case 65-12, we dealt with a situation in which a group of engineers believed that certain machinery was unsafe, and we determined that the engineers were ethically justified in refusing to participate in the processing or production of the product in question. We recognized in that case that such action by the engineers would likely lead to the loss of employment.

More recently, in Case 82-5, the engineer was employed by a large industrial company and after reviewing plans for materials supplied by a subcontractor, determined that they were inadequate both from a design and a cost standpoint and therefore should be rejected. Thereafter, the engineer advised his superiors of the deficiencies but his recommendations were rejected. The engineer persisted with his recommendations and was placed on probation with the warning that if his job performance did not improve he would be terminated.

In finding that an engineer does not have an ethical obligation to continue an effort to secure a change in the policy of an employer under these circumstances, or to report his concerns to the proper authority, we stated, nevertheless, that the engineer has an ethical "right" to do so as a matter of personal conscience. We emphasized, however, that the case then before us did not directly involve the protection of the public safety, health, and welfare, but rather was an internal dispute between an employer and an employee.

In addition, we found in Case 82-5 that the situation presented has become well known in recent years as "whistleblowing" and if an engineer feels strongly that an employer's course of action is improper when it relates to public concerns, and if the engineer feels compelled to "blow the whistle" to expose the facts as he sees them, he may well have to pay the price of loss of employment. We also commented that in recent years, engineers have gone through such experiences and even if they have ultimately prevailed on legal or political grounds, the experience is not to be taken lightly. We concluded that "the *Code* only requires that the engineer withdraw from a project and report to proper authorities when the circumstances involve endangerment to the public safety, health and welfare."

Clearly, the case presently before the Board involves "endangerment to the public safety, health and welfare"—the contamination of the water supply—and therefore it is clear that Engineer A has an obligation to report the matter to her employer. Under the facts it appears that Engineer A has fulfilled this specific aspect of her obligation by reporting her concerns to City Administrator C and thereafter to certain members of the city council. However, under the facts of this case, we believe Engineer A had an ethical obligation under the *Code* to go considerably farther.

As noted in Case 82-5 and in the *Code*, where an engineer determines that a case may involve a danger to the public safety, the engineer has not merely an "ethical right" but has an "ethical obligation" to report the matter to the proper authorities and withdraw from further service on the project. We believe this is particularly clear when the engineer involved is a public servant (city engineer and director of public works).

In the context of this case, we do not believe that Engineer A's act of reporting her concerns to City Administrator C or certain members of the city council constituted a reporting to the "proper authorities" as intended under the *Code*.

Nor do we believe Engineer A's decision to assume no responsibility for the plant and beds constitutes a "withdrawal from further service on the project."

It is clear under the facts of this case that Engineer A was aware of a pattern of ongoing disregard for the law by her immediate superior as well as members of the city council. After several attempts to modify the views of her superiors, it is our view that Engineer A knew or should have known that the "proper authorities" were not the city officials, but more probably state officials (i.e., state water pollution control authority). We cannot find it credible that a City Engineer/Director of Public Works for a medium-sized town would not be aware of this basic obligation. Engineer A's inaction permitted a serious violation of the law to continue and appeared to make Engineer A an "accessory" to the actions of City Administrator C and the others.

It is difficult for us to say exactly at what point Engineer A should have reported her concerns to the "appropriate authorities." However, we would suggest that such reporting should have occurred at such time as Engineer A was reasonably certain that no action would be taken concerning her recommendations either by City Administrator C or the members of the city council and, that in her professional judgment, a probable danger to the public safety and health then existed.

In addition, we find it troubling that Engineer A would permit her professional integrity to be compromised in the manner herein described. As the legally established city engineer and director of public works, Engineer A allowed her engineering authority to be circumvented and overruled by a nonengineer under circumstances involving the public safety. It is clear that Engineer A had an ethical obligation to report this occurrence to the "proper authorities" as stated above.

In closing, we must acknowledge a basic reality that must confront all engineers faced with similar decisions. As we noted in Cases 65-12 and 82-5, the engineer who makes the decision to "blow the whistle" will in many instances be faced with the loss of employment. While we recognize this sobering fact, we would be ignoring our obligation to the *Code* and hence to the engineering profession if, in matters of public health and safety, we were to decide otherwise. For an engineer to permit her professional obligations and duties to be compromised to the point of endangering the public safety and health does grave damage to the image and interests of all engineers.

Conclusion

Engineer A did not fulfill her ethical obligations by informing the City Administrator and certain members of the city council of her concerns.

Using Another Engineer's Proposal (BER Case 83-3)

Facts

Engineer B submitted a proposal to a county council following an interview concerning a project. The proposal included technical information and data that the council requested as a basis for the selection. Smith, a staff member of the council, made Engineer B's proposal available to Engineer A. Engineer A used Engineer B's proposal without Engineer B's consent in developing another proposal, which was subsequently submitted to the council. The extent to which Engineer A used Engineer B's information and data is in dispute between the parties.

Question

Was it unethical for Engineer A to use Engineer B's proposal without Engineer B's consent in order for Engineer A to develop a proposal which Engineer A subsequently submitted to the council?

Discussion

The Board of Ethical Review operates on an "ad hoc" educational basis, and does not engage in resolving disputes of fact between parties in actual cases. That function is left to the state society if members are involved in judging whether a member has violated the *Code of Ethics*. Being solely educational, the function of the Board of Ethical Review is to take the submission of "facts" as the basis for analysis and opinion without attempting to obtain rebuttal or comment from other parties. On that basis, the reader of the opinions should always recognize that the Board of Ethical Review is not an adjudicatory body, but its opinions are intended to apply to actual cases only to the extent of the "facts" stated in the case.

This case presents a series of facts, some of which may be addressed by the Board of Ethical Review, others that may not. It appears from the facts that a nonengineer committed certain wrongdoings. However, the Board of Ethical Review does not review the conduct of nonengineers with respect to the *Code of Ethics*. Nonengineers, of course, are not covered by the *Code* and therefore it would be a meaningless act for this Board to review the conduct of Smith in the facts presented above. Instead, it is the duty of the Board to focus upon the actions of Engineer A.

In Case 64-7, the Board interpreted Section III.9.a. (then Sections 14 and 14(a)) to mean that individual accomplishments and the assumption of responsibility by individual engineers should be recognized by other engineers. "This principle," said the Board, "is not only fair and in the best interests of the profession, but it also recognizes that the professional engineer must assume personal responsibility for decisions and actions." Although the facts of that case were somewhat different from those in the case at hand, Case 64-7 reflects the view that each individual engineer has an ethical obligation to recognize and give credit to the creative products of other engineers. At a bare minimum, that ethical obligation includes securing the consent of that engineer, indicating on any reproduction of that creation the identity of the engineer and in some cases providing the engineer with remuneration for his work depending upon the surrounding circumstances. Each case must be decided upon its individual facts, as no two cases are alike. However, certain basic obligations exist that must be recognized in all cases.

If in fact Engineer A used the proposal, it is clear that such a use would be in violation of Section III.9.a. of the *Code of Ethics*. Although it may be argued that the *Code* provision is meant to address those situations where a supervising engineer fails to give credit to an employee responsible for a particular design, and not where "proposals" (which might in fact even be a matter of public record) are submitted by several firms and one engineer merely reviews another set of proposals to gain another firm's perspective of the project, we are convinced that the *Code* may properly be read to imply use and thus proscribe the conduct of Engineer A. The Board concludes from the facts that the general purpose of Engineer A's use of the proposal of Engineer B was to develop a proposal and thus be awarded the contract. That being the purpose, Engineer A had an obligation (1) to seek and obtain Engineer B's consent before using the plans as a basis for one's own proposal; (2) if granted consent, to identify Engineer B in all cases of use of Engineer B's proposal; and (3) to negotiate and pay Engineer B "fair and reasonable" compensation for using the proposal. By failing to fulfill any of those obligations, Engineer A clearly violated Sections III.9. and III.9.a. of the *Code*.

The actions of Engineer A suggest conduct unbecoming of a professional engineer. When offered the contents of Engineer B's proposal by Smith, Engineer A had an ethical obligation to refuse to accept the proposal. Instead, Engineer A accepted and also used the material. Because of the decision to actually use the material, we must further conclude that Engineer A violated Section III.6. of the *Code* by competing unfairly with Engineer B by attempting to "obtain...advancement...by...improper or questionable methods." Although that *Code* provision is broad and leaves a good deal of room for interpretation, we are convinced that the use of the proposal constituted unfair competition by improper and questionable methods. Whether there would have been a violation of Section III.6. had Engineer A not used Engineer B's proposal but merely reviewed it before developing the proposal is a debatable point that we will leave for another day. However, this Board is being asked to determine whether a violation occurred as a result of Engineer A's use of Engineer B's proposal. We think that Engineer A's use under the present facts constitutes unfair competition by improper and questionable methods and hence a violation of Section III.6. of the *Code*.

Conclusion

It was unethical for Engineer A to use Engineer B's proposal without Engineer B's consent in order to develop a proposal that was subsequently submitted to the council.

Design Dispute (BER Case 84-4)

Facts

Client hires Engineer A to design a particular project. Engineer A develops what he believes to be the best design and meets with the client to discuss the design. After discussing the design plans and specifications, the client and Engineer A are involved in a dispute concerning the ultimate success of the project. The client believes Engineer A's design is too large and complex and seeks a simpler solution to the project. Engineer A believes a simpler solution will not achieve the result and could endanger the public. The client demands that Engineer A deliver over to him the drawings so that he can present them to Engineer B to assist Engineer B in completing the project to his liking. The client is willing to pay for the drawings, plans, specifications, and preparation but will not pay until Engineer A delivers over the drawings. Engineer A refuses to deliver the drawings.

Question

Would it be ethical for Engineer A to deliver over the plans and specifications to the client?

Discussion

The facts of the case presented to the Board, at first glance, appear to be fairly straightforward and easily addressed by the *Code of Ethics*. On its face we are presented with an engineer who has been retained by a client to design a project. However, both parties cannot agree as to the ultimate success of the project as developed by Engineer A. Thus, the client seeks to terminate the services of Engineer A, but wishes to obtain the drawings, plans, and specifications from Engineer A for a fee. Our discussion will be limited to the ethical rather than the contractual considerations of this case.

Much of the language contained in the *Code* relates to the engineer's obligation to protect the public health, property, and welfare (Section II.1.a.). In the present case it appears that Engineer A had a strong concern for the protection of the public health and welfare. Nevertheless, it is the view of this Board that Engineer A could have delivered over the drawings to the client and his conduct would have been ethically proper.

While it is true that Engineer A has an ethical obligation under Section II.1.a., that obligation assumes that Engineer A is in possession of verifiable facts or evidence which would substantiate a charge that an actual danger to the public health or safety exists. In the instant case, Engineer A makes the overly broad assumption that if he were to deliver over to the client the drawings so that the client can present them to Engineer B to assist Engineer B in completing the project to the client's liking, Engineer B would develop a set of plans which would endanger the public health and safety. We think that such an assumption is ill-founded and is not based upon anything more than a supposition by Engineer A. Therefore, we are of the view that Engineer A should not have withheld the drawings on the basis of Section II.1.a.

In reviewing the conduct of Engineer A up until his refusal to deliver over the drawings to the client, we are of the view that Engineer A went as far as he was ethically required to go in preparing what he believed was the best design for the project and in informing the client of the dangers of proceeding with the client's simplified solution. Section III.1.b. is very clear in stating an "Engineer shall advise [his] client . . . when [he] believes a project will not be successful." We are of the view that, by conferring with the client and explaining his concerns over a proposed simplified solution, Engineer A had met his ethical responsibility.

In the event, however, that Engineer A does deliver over to the client the plans so that the client can present them to Engineer B for completion of the project to the client's liking, and thereafter Engineer A discovers that Engineer B developed plans which constitute a danger to the public, certain actions would then be required by Engineer A under the *Code*. Any verifiable conduct on the part of Engineer B which indicates that Engineer B's plans are a danger to the public, should be brought to the attention of the proper authorities, i.e., the responsible professional societies or the state engineering registration board.

Conclusion

It would be ethical under the above circumstances for Engineer A to deliver over the plans and specifications to the client.

Public Criticism by Engineers (BER Case 88-7)

Facts

Engineer A, a renowned structural engineer, is hired for a nominal sum by a large city newspaper to visit the site of a state bridge construction project, which has had a troubled history of construction delays, cost increases, and litigation primarily as a result of several well publicized, on-site accidents. Recently the state highway department has announced the date for the opening of the bridge. State engineers have been proceeding with repairs based upon a specific schedule.

Engineer A visits the bridge and performs a one-day visual observation. Her report identifies, in very general terms, potential problems and proposes additional testing and other possible engineering solutions. Thereafter, in a series of feature articles based upon information gleaned from Engineer A's report, the newspaper alleges that the bridge has major safety problems that jeopardize its successful completion date. Allegations of misconduct and incompetence are made against the project engineers and the contractors as well as the state highway department. During an investigation by the state, Engineer A states that her report was intended merely to identify what she viewed were potential problems with the safety of the bridge and was not intended to be conclusive as to the safety of the bridge.

Question

Was it ethical for Engineer A to agree to perform an investigation for the newspaper in the manner stated?

Discussion

The technical expertise that engineers can offer in the discussion of public issues is vital to the interests of the public. We have long encouraged engineers to become active and involved in matters concerning the well being of the public. Moreover, the NSPE *Code of Ethics* makes clear that engineers should "seek opportunities to be of constructive service in civic affairs and work for the advancement of the safety, health and well-being of their community" (Section III.2.a.).

Obviously, this important involvement must be appropriate to the circumstance of the situation. In situations where an engineer is being asked to provide technical expertise to the public discussion, the engineer should offer objective, truthful, and dispassionate professional advice that is pertinent and relevant to the points at issue. The engineer should only render a professional opinion publicly, when that opinion is (1) based upon adequate knowledge of the facts and circumstance involved, and (2) the engineer clearly possesses the expertise to render such an opinion.

The Board has earlier visited situations in which engineers have publicly rendered professional opinions. In Case 65-9, a consulting engineer who had performed the engineering work on a portion of an interstate highway to which a proposed controversial highway by-pass would connect, issued a public letter which was published in the local press, criticizing the cost estimates of the engineers of the state highway department, stating alleged disadvantages of the proposed route, and pointing out an alternative route. The newspaper story contained the full text of the letter from the consulting engineer.

In deciding that it was ethical for the engineer to publicly express criticism of the proposed highway routes prepared by engineers of the state highway department, the Board stated: "...the whole purpose of engineering is to serve the public interest. When an engineering project has such a direct and substantial impact on the daily life of the citizenry as the location of a highway it is desirable that there is public discussion. The *Code* does not preclude engineers, as citizens, from participating in such public discussion. Those engineers who have a particular qualification in the field of engineering involved may be said to even have a responsibility to present public comment and suggestions in line with the philosophy expressed in the *Code*."

Thereafter, in Case 79-2, the Board ruled that where an engineer had significant environmental concerns, it was not unethical for the engineer to criticize a town engineer and a consulting engineer with respect to findings contained in a report on a sanitary landfill for the town. Said the Board: "It is axiomatic that an engineer's primary ethical responsibility is to follow the mandate of the *Code* to place the public welfare over all other considerations."

We noted that these issues in the public arena are subject to open public debate and resolution by appropriate public authority. Here the engineer was acting within the intent of the *Code* in raising his concern. We concluded by citing earlier decision Case 63-6 in which we noted: "There may also be honest differences of opinion among equally qualified engineers on the interpretation of the known physical facts....The *Code* does not prohibit...public criticism; it only requires that the engineer apply due restraint...in offering public criticism of the work of another engineer; the engineering witness will avoid personalities and abuse, and will base his criticism on the engineering conclusions or application of engineering data by offering alternative conclusions or analyses." It is clear, based upon the *Code of Ethics* and several interpretations of the *Code* by this Board, that the engineer may and, indeed in some cases, must ethically provide technical judgment on a matter of public importance with the aforementioned considerations concerning expertise, adequacy of knowledge, and the avoidance of personality conflicts in mind.

However, we must note that under the facts of this case, we are not merely dealing with a disinterested engineer who on her own has decided to come forward and offer her professional views. Rather, we are dealing with an engineer who was retained by a newspaper to provide her professional opinion with the understanding that the opinion could serve as the basis for news articles concerning the safety of the bridge. This fact gives an added ethical dimension to the case and requires our additional analysis. In this regard, it is our view that as a condition of her retention by the newspaper involved, Engineer A has an ethical obligation to require that the newspaper clearly state in the articles that Engineer A had been retained for a fee by the newspaper in question to perform the one-day observation of the bridge site.

We should also add that in circumstances such as here where an engineer is being retained by a newspaper to offer a professional opinion concerning a matter of public concern, the engineer must act with particular care, should exercise the utmost integrity and dignity, and should take whatever reasonable steps are necessary to enhance the probability that the engineer's professional opinions are reported completely, accurately, and not out of context. While we recognize that there are limits to what an engineer can do in these areas, we believe that the engineer has an obligation to the public as well as to the profession to protect the integrity of her professional opinions and the manner in which those opinions are disseminated to the public.

Conclusion

It was not unethical for Engineer A to agree to perform an investigation for the newspaper in the manner stated but Engineer A has an obligation to require the newspaper to state in the article that Engineer A had been retained for a fee by the newspaper to provide her professional opinion concerning the safety of the bridge.

Withholding Information Affecting Public Safety (BER Case 76-4)

Facts

The XYZ Corporation has been advised by a State Pollution Control Authority that it has 60 days to apply for a permit to discharge manufacturing wastes into a receiving body of water. XYZ is also advised of the minimum standard that must be met.

In an effort to convince the authority that the receiving body of water after receiving the manufacturing wastes will still meet established environmental standards, the corporation employs Engineer Doe to perform consulting engineering services and submit a detailed report.

After completion of his studies but before completion of any written report, Doe concludes that the discharge from the plant will lower the quality of the receiving body of water below established standards. He further concludes that corrective action will be very costly. Doe verbally advises the XYZ Corporation of his findings. Subsequently, the corporation terminates the contract with Doe with full payment for services performed, and instructs Doe not to render a written report to the corporation.

Thereafter, Doe learns that the authority has called a public hearing and that the XYZ Corporation has presented data to support its view that the present discharge meets minimum standards.

Question

Does Doe have an ethical obligation to report his findings to the authority upon learning of the hearing?

Discussion

Section II.4. of the *Code* is clear in providing that the engineer "will act in professional matters for each client or employer as a faithful agent or trustee." In this spirit Engineer Doe has advised the XYZ Corporation that the results of his studies indicate that the established standards will in his opinion be violated. His verbal advice to the corporation would seem to meet the letter and spirit of Sections II.4. and III.1.b.

The termination of Doe's contract with full payment for services rendered is a business decision which we will presume is permitted by the terms of the engineering services contract between Doe and his client. Doe, however, has reason to question why the corporation specifically stipulates that he not render a written report. Upon learning of the hearing, he is squarely confronted with his obligations to the public concerning its safety, health, and welfare. Section II.1. requires that his duty to the public be paramount. In this case, it is presumed that a failure to meet the minimum standards established by law is detrimental to the public health and safety.

We note that we have not heretofore during the entire existence of the board had occasion to interpret Section III.2.b. of the *Code*. That portion of Section III.2.b. which requires the engineer to report any request for "unprofessional" conduct to "proper authorities" is particularly pertinent in the situation before us. The client's action instructing Doe to not render a written report when coupled with XYZ's testimony at the hearing raises the question of Doe's obligation under Section III.2.b. We interpret the language in the context of the facts to mean that it would now be "unprofessional conduct" for Doe to not take further action to protect the public interest.

It is not material, in our view, that the subject matter does not involve plans and specifications as stipulated in Section III.2.b. We interpret "plans and specifications" in this section to include all engineering instruments of service. That particular reference must be read in light of the overall thrust of Sections II.1. and II.1.a., both of which indicate clearly that the paramount duty of the engineer is to protect the public safety, health and welfare in a broad context. As we noted in Case No. 67-10, even though involving unrelated facts and circumstances, "It is basic to the entire concept of a profession that its members will devote their interests to the public welfare, as is made abundantly clear in Section II.1. and Section II.1.a. of the *Code*."

Section III.4. of the *Code* does not give us pause because the action of the engineer in advising proper authority of the apparent danger to the public interest will not in this case be disclosing the technical processes or business affairs of the client.

Conclusion

Doe has an ethical obligation to report his findings to the authority upon learning of the hearing.

Reporting Safety Violations (BER Case 89-7)

Facts

Engineer A is retained to investigate the structural integrity of a 60-year-old occupied apartment building which his client is planning to sell. Under the terms of the agreement with the client, the structural report written by Engineer A is to remain confidential. In addition, the client makes clear to Engineer A that the building is being sold "as is" and he is not planning to take any remedial action to repair or renovate any system within the building prior to its sale.

Engineer A performs several structural tests on the building and determines that the building is structurally sound. However, during the course of providing services, the client confides in Engineer A and informs him that the building contains deficiencies in the electrical and mechanical systems which violate applicable codes and standards. While Engineer A is not an electrical nor mechanical engineer, he does realize those deficiencies could cause injury to the occupants of the building and so informs the client.

In his report, Engineer A makes a brief mention of his conversation with the client concerning the deficiencies: however, in view of the terms of the agreement, Engineer A does not report the safety violations to any third party.

Question

Was it ethical for Engineer A not to report the safety violations to the appropriate public authorities?

Discussion

The facts presented in this case raise a conflict between two basic ethical obligations of an engineer: The obligation of the engineer to be faithful to the client and not to disclose confidential information concerning the business affairs of a client without that client's consent, and the obligation of the engineer to hold paramount the public health and safety.

Section III.4. can be clearly understood to mean that an engineer has an ethical obligation not to disclose confidential information concerning the business affairs of any present client without the consent of that client. That provision makes no specific exception to the language. For example, the drafters of the *Code* could have provided exceptional circumstances where such confidential information could be disclosed by the engineer; however no such provisions have been included.

There are various rationales for the nondisclosure language contained in the *Code*. Engineers, in the performance of their professional services, act as "agents" or "trustees" to their clients. They are privy to a great deal of information and background concerning the business affairs of their client. The disclosure of confidential information could be quite detrimental to the interests of their client and therefore engineers as "agents" or "trustees" of such information are expected to maintain the confidential nature of the information revealed to them in the course of rendering their professional services.

On numerous occasions, this Board has interpreted the language contained in Sections II.4. and III.4. particularly in the context of the obligations of employed engineers to maintain the confidences of their employer particularly with regard to certain confidential information which might be made available to the engineer during the course of employment as in Case 61-8. However, more recently, the Board has interpreted this language in the context of the relationships owed by the engineer in private practice to the client.

For example, in Case 82-2, an engineering consultant performed home inspection services for a prospective purchaser of a residence and thereafter disclosed the contents of the report to the real estate firm handling the sale of the residence. The Board reaffirmed the principle of the right of confidentiality on behalf of the client. There we noted that this was not a case of an engineer allegedly violating the mandate of Section III.4. not to disclose information concerning the business affairs of the client. We said that Section III.4. necessarily relates to confidential information given the engineer by the client in the course of providing services to the client. In Case 82-2, there was no transmission of confidential information by the client to the engineer.

However, under the facts of the present case, there was a transmission of confidential information by the client to Engineer A. Therefore, it would appear that Section III.4. should be involved in our consideration of this case.

While we noted earlier that the *Code* makes no direct exception to the language contained in Section III.4., as we have stated on numerous occasions, no section of the *Code* should be read in a vacuum or independent of the other provisions of the *Code*. Section II. l .c. provides additional guidance in this case making it clear that the Engineer A has an ethical obligation to refrain from disclosing information which he acquires during the course of providing professional services to the client unless first obtaining the client's consent to disclose. Importantly, however, this section also includes a relevant exception which allows the engineer to disclose information acquired during the course of providing professional services to the client if such disclosure is authorized or required by law or by the *Code*. In other words if the engineer has a legal or ethical responsibility to disclose the information in question, the engineer is released from the obligation to maintain confidentiality.

The Board has interpreted Section II.l.c. on three different occasions (Cases 82-2, 85-4, and 87-2) but in none of those cases has the Board outlined the scope of the *Code* section. Based upon previous BER cases, it is not clear which provisions of the law would have to be involved before an engineer is released from the obligation of nondisclosure. Similarly and more importantly, from a review of these cases for our purposes, it is unclear which provisions of the *Code of Ethics* would apply before an engineer is released from the obligation of nondisclosure.

However, we believe we can draw from another case involving somewhat different facts to illustrate an approach which might be applicable to the case at hand. Case 84-5 involved a client who planned a project and hired an engineer to furnish complete engineering services for the project. Because of the potentially dangerous nature of implementing the design during the construction phase, the engineer recommended that a full-time, on-site project representative be hired for the project. After reviewing the complete project plans and costs, the client indicated to the engineer that the project would be too costly if such a representative were hired. They found it was unethical for the engineer to proceed with his work on the project knowing that the client would not agree to hire a full-time project representative. The Board noted that the engineer acceded to the client's wishes and proceeded with the work despite the fact that the engineer believed that to proceed without an on-site project representative would be potentially dangerous. The engineer did not force the issue or insist that a project representative be hired. Instead, the engineer "went along" without dissent or comment. If the engineer's ethical concerns were real, the engineer should have insisted that the client hire the on-site project representative or refuse to continue to work on the project.

We believe much of the same reasoning applies in the present case. Under the reasoning of Case 84-5, the engineer had an obligation to go further. We believe under the facts, Section II.l.c. should be read in conjunction with Section II.l.a. The latter section refers to the primary obligation of the engineer to protect the safety, health, property, and welfare of the public. The obligation of the engineer to refrain from revealing confidential information, data, and facts concerning the business affairs of the client without consent of the client is a significant ethical obligation. We further believe that matters of public health and safety must take precedence. The *Code of Ethics* is clear on this point. Section I.1. employs the word "paramount" to describe the obligation of the engineer with respect to public health and safety.

We believe Engineer A could have taken other steps to address the situation, not the least of which was his paramount professional obligation to notify the appropriate authority if his professional judgment is overruled under circumstances where the safety of the public is endangered. Instead, Engineer A, like the engineer in Case 84-5, "went along" and proceeded with the work on behalf of the client. His conduct cannot be condoned under the *Code*.

Conclusion

It was unethical for Engineer A not to report the safety violations to the appropriate public authorities.

Failure to Report Safety Defects (BER Case 90-5)

Facts

Tenants of an apartment building sue the owner to force him to repair many defects in the building which affect the quality of use. Owner's attorney hires Engineer A to inspect the building and give expert testimony in support of the owner. Engineer A discovers serious structural defects in the building which he believes constitute an immediate threat to the safety of the tenants. The tenants' suit has not mentioned these safety related defects. Upon reporting the findings to the attorney, Engineer A is told he must maintain this information as confidential as it is part of a lawsuit. Engineer A complies with the request of the attorney.

Question

Was it ethical for Engineer A to conceal his knowledge of the safety-related defects in view of the fact that it was an attorney who told him he was legally bound to maintain confidentiality?

Discussion

The obligation of the engineer to protect the public health and safety has long been acknowledged by the *Code of Ethics* and by the Board of Ethical Review. This responsibility rests with the recognition that engineers with their education, training and experiences possess a level of knowledge and understanding concerning technical matters which is superior to that of the lay public. It also is rooted in the implicit fact that as individuals who are granted a license by the state to practice, engineers have a duty to engage in practice which is consistent with the interests of the state and its citizenry.

This board has long recognized this obligation. A good example is BER Case 84-5. There, a client planned a project and hired Engineer A to furnish complete engineering services for a project. Because of the potentially dangerous nature of implementing the design during the construction phase, Engineer A recommended to the client that a full-time, on-site project representative should be hired for the project. After reviewing the completed project plans and costs, the client indicated to Engineer A that the project would be too costly if such a representative were hired.

Engineer A proceeded with the work on the project even though he had recommended that a full-time, on-site project representative should be hired. In discussing the issue of whether it was unethical for Engineer A to proceed with work on the project knowing that the client would not agree to hire a full-time, on-site project representative, the Board noted that Section II.I.a. admonishes engineers to recognize that their primary obligation is to protect the public safety, health, property and welfare. Under the facts, Engineer A did not recognize this primary obligation. Engineer A, using his best professional judgment, made a recommendation consistent with that obligation. However, when costs concerns where raised by the client, Engineer A abandoned the ethical duty and proceeded with the work on the project. The Board concluded that Engineer A appeared to have acted in a manner that suggests that the primary obligation was not to the public but to the client's economic concerns. For that reason, Engineer A was in violation of Section II.1.a. of the *Code*.

Although the public health and safety clearly is the most basic and fundamental ethical obligation of engineers, other important ethical obligations exist for which engineers must be ever mindful. One important ethical consideration is the obligation of engineers not to reveal information of the client without the prior consent of the client.

The Board has had reason to consider this ethical issue on occasion. In BER Case 82-2, Engineer A offered home inspection services, whereby Engineer A undertook to perform an engineering inspection of residences by prospective purchasers. Following an inspection, Engineer A would render a written report to the prospective purchaser. Engineer A performed this service for a client for a fee and prepared a one-page written report, concluding that the residence was in generally good condition requiring no major repairs, but noting several minor items needing attention.

Engineer A submitted his report to the client showing that a carbon copy was sent to the real estate firm handling the sale of the residence. The client objected that such action prejudiced their interests by lessening their bargaining position with the owners of the residence. They also complained that Engineer A acted unethically in submitting a copy of the report to others who had not been a party to the agreement for the inspection services. In concluding that Engineer A acted unethically in submitting a copy of the home inspection to the real estate firm representing the owner, the Board concluded that although it did not appear from the facts that Engineer A had acted with some ulterior motive or intention to cause the client any harm, the principle of the right of confidentiality on behalf of the client predominated.

Given these two cases, it is clear that there may be facts and circumstances in which the ethical obligation of engineers in protecting the public health and safety conflict with the ethical obligation of engineers to maintain the right of confidentiality in data and other information obtained on behalf of a client. While we recognize that this conflict is a natural tension which exists within the *Code*, we think that under the facts of this case, there were reasonable alternatives available to Engineer A which could assist him in averting an ethical conflict.

It appears that Engineer A, having become aware of the imminent danger to the structure, had an obligation to make absolutely certain that the tenants and public authorities were made immediately aware of the dangers that existed. Engineer A's client was the attorney and technically Engineer A had an obligation not to reveal facts, data, or other information in a professional capacity without the prior consent of the attorney. However, there were valid reasons why Engineer A should have revealed the information directly to the tenants and public authorities.

Unlike the facts presented in BER Case 82-2, there is not any conflict or potential conflict of interests that exists between owner and attorney with regard to the information. Although Attorney retained Engineer A directly, he did so on behalf and for the benefit of the owner. Therefore, the key issue in BER Case 82-2 upon which an ethical violation was found, is absent in this case.

Section II.1.c. makes a clear exception concerning the obligation of engineers not to reveal facts obtained in a professional capacity without the client's consent. That exception allows the disclosure of such information in cases authorized by the *Code* or required by law. We believe that in cases where the public health and safety is endangered, engineers not only have the right but also the ethical responsibility to reveal such facts to the proper persons. We also believe that state board rules of professional conduct might require such action by professional engineers.

Conclusion

It was unethical for Engineer A not to report the information directly to the tenants and public authorities.

CONFLICT OF INTEREST CASES

Personal Interest in Public Project (BER Case 88-1)

Facts

Engineer A is retained by the county to perform a feasibility study and make recommendations concerning location of a new power facility in the county. Two parcels of land located on a river have been identified by the county as the "candidates" for facility sites. The first parcel is undeveloped and owned by an individual who plans to build a recreational home for his family. The second parcel, owned by Engineer A, is developed. Engineer A discloses that he is the owner of the second parcel of land and recommends that the county build the facility on the undeveloped parcel of land because (1) it is a better location for the power facility from an engineering standpoint, and (2) it would be less costly for the county to acquire. The county did not object to having Engineer A perform the feasibility study.

Question

Was it ethical for Engineer A to perform a feasibility study and make recommendations concerning the location of a new power facility in the county?

Discussion

The issue of conflict of interest is one of the most widely discussed and debated in engineering ethics. As a customary proposition, it is generally recognized as good practice for engineers to endeavor to avoid conflicts of interest. However, the language in the NSPE *Code of Ethics* relating to conflicts of interest has been significantly modified over the years. For many years, the NSPE *Code* contained strict proscriptions against engineers engaging in conflicts of interest and admonished engineers, in strong language, to avoid such conflicts. While the current language still maintains a strong tone, it is more general than in the past, requiring the disclosure of conflicts rather than complete avoidance.

The Board has had various occasions to interpret the language of the *Code of Ethics* addressing the subject of conflict of interest. For example, in Case 69-13, the Board reviewed a situation where an engineer was an officer in an incorporated consulting engineering firm that was primarily engaged in civil engineering projects for clients. Early in the engineer's life, he had acquired a tract of land by inheritance, which was in an area being developed for residential and industrial use. The engineer's firm had been retained to study and recommend a water and sewer system in the general area of his land interest. The question faced by the Board under those facts was, "May the engineer ethically design a water and sewer system in the general area of his land interest?" The Board ruled that the engineer could not ethically design the system under those circumstances.

The Board recognized that the issue was a difficult one to resolve, pointing to the fact that there was no conflict of interest when the engineer entered his practice. The conflict developed in the normal course of his practice when it became apparent that his study and recommendation could lead to the location of a water and sewer system that might cause a considerable appreciation in the value of his land depending upon the exact location of certain system elements in proximity to his land. The Board stated that while the engineer must make full disclosure of his personal interest to his client before proceeding with the project, such disclosure was not enough under the *Code*. The Board concluded by saying: "This is a harsh result, but so long as men are in their motivations somewhat 'lower than angels,' it is a necessary conclusion to achieve compliance with both the letter and the spirit of the *Code of Ethics*. The real test of ethical conduct is not when compliance with the *Code* comports with the interest of those it is intended to govern, but when compliance is adverse to personal interest."

More recently in Case 85-6, the Board reviewed similar facts and circumstances and came to a different result. There, an engineer was retained by the state to perform certain feasibility studies relating to a possible highway spur. The state was considering the possibility of constructing the highway spur through an area that was adjacent to a residential community in which the engineer's residence was located. After learning of the proposed location of the spur, the engineer disclosed to the state the fact that his residential property might be affected and fully disclosed the potential conflict with the state. The state did not object to the engineer performing the work. Engineer A proceeded with his feasibility study and ultimately recommended that the spur be constructed.

In ruling that it was not unethical for the engineer to perform the feasibility study despite the fact that his land might be affected thereby, the Board noted that the ethical obligations contained in Section II.4.a. do not require the engineer to "avoid" any and all situations that may or may not raise the specter of a conflict of interest. Such an interpretation of the *Code of Ethics*, said the Board, would leave engineers without any real understanding of the ethical issues nor any guidance as to how to deal with the problem. We noted that the basic purpose of a code of ethics is to provide the engineering profession with a better awareness and understanding of the ethical issues that impact upon the public. The Board concluded that only through interacting with the public and clients will engineers be able to comprehend the true dimensions of ethical issues.

While one can read Case 85-6 and possibly draw the conclusion that in the instant case Engineer A's conduct was ethically proper, we do not reach that conclusion. It is our view that the two cases should be distinguished. In Case 85-6, the benefit to be derived by the engineer from the construction of the spur in question was far more remote than the benefit in the case at hand. The construction of the highway spur presumably would add to the value of the engineer's residential property, but it would not impact upon his ownership of the property. In the instant case, Engineer A is being placed in a position whereby he is making a recommendation that could directly affect his and his neighbor's ownership in property. It is one thing for an engineer to participate in decisions that will have a tangential impact upon his interests as was the case in 85-6. It is quite another matter for the engineer to act in his professional capacity to advise a governmental entity on policy matters where his economic interests are directly at issue. We find ourselves in agreement with the reasoning contained in Case 69-13 which we believe is more applicable to the facts present here.

We are reminded that Engineer A's professional opinion was supported by two important public policy considerations. First, it was noted by Engineer A that the undeveloped parcel was a better location for a power facility from an engineering standpoint. Second, it was indicated that the cost to the county of acquiring the developed property would be higher than the cost of acquiring the undeveloped tract of land. While these two considerations are important ones from a public policy standpoint, and may even be rationalized by a perfunctory reading of Section III.1.b. of the *Code of Ethics*, they are not sufficient to justify Engineer A's decision to perform the feasibility study for the county. Public perceptions play an important role in engineering ethics. The facts and circumstances of Engineer A's study may appear to suggest a benefit to the "common good" if his recommended course of action is followed. That notwithstanding, a loss of public confidence may cause a damage that cannot be easily repaired due to the appearance of impropriety.

The far simpler and more ethical approach which we believe should have been followed by Engineer A, under the circumstances in this case, was recommended in Case 69-13 which states: "[The Engineer] can avoid such a conflict under these facts either by disposing of his land holdings prior to undertaking the commission or by declining to perform the services if it is not feasible or desirable for him to dispose of his land at the particular time."

Conclusion

It was unethical for Engineer A to perform a feasibility study and make recommendations concerning the location of a new power facility in the county.

Acceptance of Gifts (BER Case 87-5)

Facts

The ABC Pipe Company is interested in becoming known within the engineering community and, in particular, to those engineers involved in the specification of pipe in construction. ABC would like to educate engineers about the various products available in the marketplace: the advantages and disadvantages of using one type of pipe over another. ABC sends an invitation to Engineer A, as well as other engineers in a particular geographic area, announcing a one-day complimentary educational seminar to educate engineers on current technological advances in the selection and use of pipe in construction. ABC will host all refreshments, buffet luncheon during the seminar, and a cocktail reception immediately following. Engineer A agrees to attend.

Question

Was it ethical for Engineer A to attend the one-day complimentary educational seminar hosted by the ABC Pipe Company?

Discussion

Ethical concerns relating to the issue of gifts and other consideration provided by suppliers to engineers are addressed in several sections of the NSPE *Code*. Obviously, instances where gifts or other property of monetary value are exchanged between an engineer and a potential client are extremely sensitive and do require careful scrutiny to determine if such exchanges are proper. In the past, this Board has examined the question from two perspectives: (1) where an engineer provides a client with a gift or valuable consideration under circumstances that could create the appearance of seeking to influence the client's judgment, and (2) where a supplier provides the engineer with a gift or valuable consideration under circumstances that could create an appearance that the supplier was seeking to influence the engineer's judgment. The instant case relates to the latter situation.

In Case 60-9 the Board examined a situation involving certain engineering employees of an industrial firm who were in a position to recommend for or against the purchase of products used by the company. They regularly received cash gifts ranging from $25 to $100 from product salesmen. In ruling that accepting those gifts was not ethical, the Board noted that an occasional free luncheon or dinner, and a Christmas or birthday present when there is a personal relationship, are acceptable practice. On the other hand, cash payments to those in a position to influence decisions favorable or unfavorable to the giver are not in good taste and do immediately raise the suspicion of an ulterior motive.

More recently, in Case 81-4, this Board dealt with three engineers who were principals or employees of a consulting engineering firm that did an extensive amount of design work for private developers. The engineers were involved in recommending to the developers a list of contractors and suppliers to be considered for selection on a bidding list for construction of some projects. Usually, those the engineers recommended obtained most of the contracts from the developers. Over a period of years, the officers of the contractors or suppliers developed a close business and personal relationship with the engineers. From time to time, at holidays or on the engineers' birthdays, the contractors and suppliers would give them personal gifts of substantial value, such as house furnishings, recreational equipment, or gardening equipment. In finding that it was unethical for the engineers to accept those gifts, we stated that engineers should "lean over backward" to avoid acceptance of gifts from those with whom they, or their firm, do business.

At that time, the Board again noted that there may be circumstances when a gift is permissible, as stated in Case 60-9, and does not compromise the engineer's independent professional judgment.

The *Code* unequivocally states that engineers must not accept gifts or other valuable consideration from a supplier in exchange for specifying its products. (See Sections II.4.c.; III.5.b.) However, in this case we are dealing with a material supplier who is introducing information about pipe products to engineers in the community and has chosen the form of an educational seminar as its vehicle. While ABC Pipe Company will seek to present its particular products in a favorable light and point out their many advantages over others', a complimentary invitation to such a seminar would not reach the level that would raise an ethical concern. The earlier decisions and the pertinent provisions of the *Code* relate more to the circumstances in which valuable gifts are received and at least create the appearance of a "quid pro quo" or an exchange of valuable consideration for specifying the equipment. Under the facts of this case, Engineer A is accepting an opportunity to become educated on a particular topic. He would be attending the seminar with many of his colleagues, and there is no suggestion in the facts that anyone at ABC Pipe Company would personally seek to persuade Engineer A to specify its products.

We view the buffet luncheon and cocktail reception immediately following the seminar as falling within the minimal provisions noted earlier in Cases 60-9 and 81-4, and thus it would not be improper for Engineer A to participate in those activities. We note, however, that had Engineer A agreed to accept items of substantial value (e.g., travel expenses, multi-day program, resort location, etc.) our conclusion would have been quite different.

Conclusion

It was ethical for Engineer A to attend the one-day complimentary educational seminar hosted by the ABC Pipe Company.

Private Employment After Public Employment (BER Case 78-10)

Facts

Engineer A had been employed for many years by a county and had specialized during a large part of that time in evaluating the engineering aspects of county zoning petitions. Engineer A retired from public employment and opened a consulting office to engage in that type of work for private clients. He was retained by the XYZ Development Corporation, which had pending before the county zoning board a petition for a zoning variance to permit it to proceed with the development of a major housing development. In order to obtain the variance, XYZ had to demonstrate the technical feasibility of a major water system consistent with environmental impact on adjacent communities. Engineer A, while employed by the county, had been involved in the XYZ petition and had made some preliminary technical evaluations on it. Citizen groups opposing the granting of the variance have formally objected to Engineer A's being allowed to participate in further proceedings as an expert witness in support of the petition.

Question

Is it ethical for Engineer A to represent the XYZ Development Corporation on a matter on which he had done some work for the county?

Discussion

This case illustrates a question which has become an issue of considerable study and debate in recent years, often referred to as the "revolving door" issue, in which professionals employed by a unit of government leave that employment and become private practitioners to specialize in and represent clients before their previous employing agency in the particular field of specialization.

We read Section III.4.a. to refer primarily to situations in which an engineer may act to the detriment of a former employer by engaging in activities directly related to the work he had done for the former employer on a particular project or assignment. In Case 74-4, for example, we said that the thrust of Section III.4.a. "...is to protect an employer and parties having an interest in his practice from a former employee utilizing this special knowledge to their detriment...." Under that reading, we have to ask in this case who, if anyone, would be "victimized" by the action of Engineer A in now representing the interests of the XYZ company.

By virtue of his long experience and expertise gained in the subject matter while employed by the county, Engineer A has gained detailed information on the internal procedures and policies of the county zoning authorities and that was undoubtedly a factor in XYZ's decision to retain Engineer A. We may also assume that during his involvement in the work of the county zoning authorities he has gained close personal contacts with the public officials and other staff members who are involved in the decision whether to grant the variance.

In this situation the public, or at least that part of it which is concerned with the variance issue, may be regarded as an "interested party" under the wording of Section 7(a).

Taking a broader view of the fundamental issue in this case, we believe that the circumstances are such that Section III.3., even though a more general statement of ethical concern, is pertinent. Giving Engineer A the benefit of the doubt as to his motive and purpose in taking on a case for a private client in which he had been directly involved as a public employee, we are concerned that his action is of a nature which may arouse public suspicion and open the way for charges which could reflect upon the profession. While all actions of engineers cannot be judged ethically on the sole basis of what some elements of the public might think or allege, in this kind of situation there is at least enough doubt to suggest that Engineer A's action was inappropriate and should be avoided in the larger interest of protecting the profession from misunderstanding.

Conclusion

It was not unethical for Engineer A to represent the XYZ Development Corporation on a matter on which he had done some work for the county, but this type of arrangement under these circumstances should be regarded as inappropriate.

Dissenting Opinion

Any action taken by a professional engineer which would tend to throw suspicion on the integrity of an engineering decision or places the integrity of the profession in a suspect position by the public sector must be considered critically in the general light of Section III.3. Since Engineer A has been involved in a specific question involving the XYZ Development Corporation while employed by the county, his appearance on behalf of the XYZ Development Corporation in the same case at a later date could be viewed by the public as a change of position based on Engineer A's self-interest. This could readily be used to impute the integrity of Engineer A's professional standing and additionally reflect adversely on the profession as a whole. While an interpretation of Section III.3. represents a judgment factor which cannot be identified as right or wrong, in the last analysis Engineer A's action tends to sully the integrity of the profession. Thus we believe that Engineer A's action was not only inappropriate but was not ethical in fulfilling the spirit or intent of Section III.3.

Author's Note: The Board of Ethical Review was evenly split in their vote on this case.

Furnishing Limited Advice (BER Case 90-7)

Facts

Engineer A serves as a principal in a consulting firm that employs other design professionals (architects, engineers, planners, surveyors). Engineer A is also retained on a part-time basis by a municipality to furnish limited advice, perform feasibility studies, and review RFPs. His agreement with the municipality states that when Engineer A's firm is employed by a client to provide professional services, other professionals in his firm and not Engineer A will be required to prepare any private client's plans that may be submitted to the municipality for review/approval. Engineer A's firm has a similar policy stating that when one of its private clients submits plans to the municipality for review/approval, Engineer A is not permitted to work on the plans at the consulting firm. Instead, those plans are prepared and sealed by Engineers B, C, and D.

Question

Would it be ethical for Engineer A to provide engineering services under the conditions herein described?

Discussion

The issue of conflict of interest and its impact on engineers acting in both a public and private capacity has been addressed by this Board on several occasions. Engineers as possessors of technical expertise are frequently called upon by both public and private parties to render advice on complex matters with consequences for public health, safety, and welfare. For obvious reasons, the interests of the various parties involved may not necessarily coincide. Where the engineer may have duties or obligations to one or more parties on such matters, the *Code* requires that the engineer must be careful to take steps which limit or avoid such conflicts.

This Board has at numerous times been asked to interpret Section II.4.d. to provide specific guidance in cases which have been presented. One good example of such an interpretation is BER Case 82-4, where an engineer in full-time private practice was retained by a county as county engineer for a stipulated monthly fee. The engineer's duties included reviewing plats and construction drawings to determine whether they met county requirements, and make recommendations to local developers, county commissions, and the planning and zoning boards. In addition, the engineer was retained by a city as city engineer for a stipulated annual fee. His duties included making recommendations to the city council concerning the approval of completed engineering work. The engineer served as project administrator for the county airport authority and as such was responsible for formulating a plan for the continued development of an airport industrial park. Finally, the engineer was administrator of the city block grant program and as such supervised engineering work on various other projects. The engineer also was retained as a consultant by several private firms to help develop city and county project proposals.

The Board of Ethical Review, reviewing earlier cases as well as the language contained in the *Code*, ruled that it would be ethical for the engineer, who served as city engineer and county engineer for a retainer fee, to provide engineering services in a private capacity to the city or the county; ethical for the engineer, who served as a member of local boards or commissions which sometimes require the services of engineers, to provide services through his private firm to those boards and commissions; and unethical for the engineer, who served as city and county engineer for a retainer fee, to provide approval or render judgment on behalf of the city and/or county relative to projects on which the engineer had furnished services through a private client.

In reaching its decisions and finding the engineer's conduct to be ethical, the Board noted that under the facts presented, the engineer did not actually participate in "decisions" with respect to the services solicited or provided by him or his organization in a private or public engineering practice but rather reviewed, recommended, formulated, and supervised plans. The Board therefore concluded that one who serves as both city and county engineer for a retainer fee may provide engineering consulting services to the city or county. With regard to status as a member of a local board or commission, the Board noted that there was nothing in that case to suggest that the engineer had taken any action to influence decisions as administrator of the city block grant program or as project administrator of the county airport authority.

Finally, in reaching its decision finding the engineer's conduct to be unethical in rendering public judgments on projects upon which the engineer furnished services to private clients, the Board, citing earlier BER Case 67-12, said that for an engineer serving as a part-time county engineer to make a recommendation on plans prepared by the engineer in his private capacity is a "useless act." The Board noted it is basic to the *Code* that an engineer cannot wear two hats and still represent the best interests of a client. The Board cautioned at that time that, as a general matter, it would be preferable for engineers to avoid such situations.

More recently, in BER Case 85-2, an engineer, a principal in a local engineering firm, served on the board of directors of a county hospital which owned a hospital facility and which contracted with a private health care provider to manage, administer and generally operate the health care facility. Certain engineering and surveying work needed to be performed at the hospital facility and the engineer received the contract from the private health care provider to perform the work. The decision to select the engineer was made by the private health care provider's board of directors and the engineer participated in the decision.

In finding that it was unethical for the engineer to seek to contract with the private health care provider to provide engineering and surveying services, we noted that as a board member of a private health care provider, which was acting in a quasi-public capacity, the engineer, consistent with Section II.4.d., could not have ethically participated in decisions with respect to professional services provided by his firm. The engineer, a principal in an engineering firm, was serving on a body that was exercising judgment and discretion in place of a governmental body and therefore should be treated as a governmental official for purposes of a conflict of interest. By participating in the private health care provider's decision to select his firm, the engineer acted in conflict with the *Code*. We should also note that at the time BER Case 859 was decided, there was no provision in Section II.4.d. which related to engineers serving on a "quasi-governmental body." However, since that time, the *Code* has been amended to include specific reference to a "quasi governmental body."

Based upon a careful reading and review of BER Cases 82-4 and 85-2, it is clear the this Board has historically made an important distinction based upon the language in Sections II.4.d. and e. between those circumstances (1) where an engineer acts in some capacity as an advisor to a public agency and also provides professional services to the public agency and (2) where an engineer is part of the decision-making group within a public agency and also providing professional services to the public agency. The former will generally, in the absence of other circumstances and factors, be found acceptable under the *Code* while the latter, on its face will be found to be unacceptable under the *Code*.

Sections II.4.d. and e. must be read together in light of earlier decisions by the Board. In particular, the prohibitions contained in Sections II.4.d. and e. would appear to apply only to those cases where an engineer is actually part of the final decision-making group of the governmental body or is acting as an officially designated, elected, or appointed member of a governmental body, and not to circumstances where, as here, the engineer is retained merely in a consultative role to a governmental body.

The Board is troubled that there is the potential for a conflict of interest should problems arise during the course of a project. The Board recognizes there may be some communities where professionals are not locally available to provide services to the client. That fact notwithstanding, the Board would recommend a contract provision under which the parties would utilize outside consultants for dispute resolution in the event a conflict of interest arises.

Conclusion

It would not be unethical for Engineer A to provide engineering services under the circumstances herein described.

ETHICAL TRADE PRACTICES

Using Designs of Others (BER Case 93-1)

Facts - Copycat Case 1

Engineer A, a registered professional engineer, has worked on the design and development of improved wastewater treatment processes and equipment, which are subsequently patented. Engineer B, an environmental consultant specializing in the design of wastewater treatment facilities, and his client are impressed with the new processes and equipment. However, Engineer B dislikes specifying sole source and, in fact, makes a point of encouraging competition by preparing open specifications with "or equal" clauses or by specifying a performance requirement. The primary, if not the sole, purpose of Engineer B's effort is to minimize cost by promoting competition. On this project, Engineer B prepares a performance specification for open competition but patterned from the performance of the processes and equipment patented by Engineer A.

Facts - Copycat Case 2

Engineer X, a registered professional engineer, has worked on the design and development of improved wastewater treatment processes and equipment which are subsequently patented. Engineer Y, an environmental consultant specializing in the design of wastewater treatment facilities, and his client are impressed with the new processes and equipment. However, Engineer Y dislikes specifying sole source. To promote competition in this instance, Engineer Y contacts several manufacturers to encourage them to develop processes and equipment that will accomplish the same results as those of Engineer X. Engineer Y provides them with proposed performance specifications patterned from those of Engineer X's processes and, as an inducement, makes a verbal commitment to include their products among "or equals" in his future specifications.

Questions

1. Is it ethical for Engineer B to use Engineer A's patented processes and equipment as a guide in preparing open specifications in order to minimize cost and to promote competition?
2. Is it ethical for Engineer Y to induce other manufacturers to produce a process and equipment that will perform equally to patented products by making verbal commitments?

General Discussion

There is concern expressed within engineering circles that the number of U.S. patents being filed by U.S. manufacturers has been declining in the past decade. As recently noted by the president of a leading patent, trademark, and copyright association, "fewer than half of U.S. patents issued will be going to U.S. inventors and the American economy will be the big loser." One alleged reason for the decline is the increase in "copycat" versions of products and processes being manufactured. One engineer recently noted that while a "better mouse trap" is terrific, the more competitively priced "copycat" versions discourage or eliminate those firms that made the investment in creating a "better mousetrap" by preventing them from recouping their original costs.

It would seem that the fundamental issue involved in these two circumstances is whether and to what extent one engineer has an ethical responsibility not to encourage others to develop alternatives based upon the technical ideas and developments of another engineer.

The Board of Ethical Review has never squarely addressed the question raised by the facts in these two cases. However, as early as BER Case 64-7, the Board noted that individual accomplishments and the assumption of responsibility by individual engineers should be recognized by other engineers. "This principle," said the Board, "is not only fair and in the best interests of the profession, but it also recognizes that the professional engineer must assume personal responsibility for decisions and actions." While BER Case 64-7 reflected the basic view that each individual engineer has an ethical obligation to recognize and give credit to the creative products of other engineer, the case did not address the question of one engineer's ethical responsibility not to persuade manufacturers to produce optional devices using the concepts of another engineer.

Over the years, the two cases that have probably come closest to addressing these issues, however remotely, are BER Cases 77-5 and 83-3. In BER Case 77-5, an engineering firm submitted a project study originally prepared for a federal client to a state agency to assist the agency in obtaining funding for a project. After obtaining the funding, the agency distributed the study to another engineering firm that used the contents of the study as part of its negotiations with the state agency. In concluding that it was ethical for the second firm to enter into negotiations for the project under the circumstances, the Board could not find any specific provisions of the *Code* which dealt either directly or indirectly with the obligations of an engineer on behalf of or as an agent of the owner to avoid taking advantage of another engineer who had in good faith provided substantial and valuable information for a proposed project on an understanding that the engineer providing the assistance would receive the commission for it. The Board, deploring the lack of specificity in the *Code*, suggested that consideration be given to an appropriate revision or addition to the *Code* to cover such a situation. Said the Board, "our reluctant conclusion may meanwhile serve the purpose of alerting engineers in private practice who are tempted to expend substantial time, effort, and funds to secure a commission to the danger they run when that investment exceeds a nominal investment."

Following the rendering of BER Case 77-5, the National Society of Professional Engineers Board of Directors took steps to modify the provisions of then Section 11 of the *Code of Ethics*. However, instead of strengthening Section 11 as recommended by the Board of Ethical Review, the Board of Directors deleted several provisions of that section in order to comply with the federal antitrust laws. An abridged version of Section 11 ultimately became the current Section III.6. contained in the *Code of Ethics*.

Later, in BER Case 83-3, which involved facts similar to BER Case 77-5, the Board concluded that it was unethical for one engineer to use the data of another engineer to develop a proposal submitted to a public authority without consent.

Case 1 Discussion

It is clear under the facts that Engineer A has devoted a great deal of time, effort and creativity to the development of the improved wastewater treatment processes and equipment. The fact that Engineer A's achievements have been granted patents is a clear demonstrations of the quality and distinction to which his work has been recognized. It may seem to some that in fairness Engineer A would be entitled to exclusive control over the fruits of his creative work and that competitors would be excluded from using the concepts and theories behind his creations to develop alternative processes and equipment that might achieve a same or similar result. However, we believe that such a notion would be inconsistent with basic principles of law as well as the philosophy which obligates engineers to cooperate in extending the effectiveness of the profession by interchanging information and experience with other engineers.

It should be noted that a fundamentally accepted principle is that an "idea," "thought," "notion," or similar abstraction cannot receive legal or other proprietary protection under the law. Rather, it is the expression of that idea, thought, or notion that can receive appropriate legal protection. This view is grounded in the philosophy that in order to best promote scientific and technological advances within our society, individuals and groups of individuals should be free to use ideas and concepts to develop different expressions of those ideas without legal hindrance. It is consistent with the principles of total quality management including the goal of constant improvement in the design process.

We believe that this basic philosophy is applicable to Case 1. Engineer B did not seek to infringe upon the patent of Engineer A. Instead, it appears that under the facts, Engineer B merely used the processes and equipment developed by Engineer A as a "standard" by which different processes and equipment would be evaluated or, as an alternative, established a performance specification based on the performance of Engineer A's processes and equipment which would be used to evaluate the performance of different processes and equipment.

As to the question of the primary or sole purpose of Engineer B's efforts (minimizing cost to client by promoting competition among suppliers), we believe such an objective is entirely consistent with the engineer's general obligation to the client to act as faithful agents or trustees (Section II.4.).

Case 2 Discussion

The facts in Case 2 are different than those involved in Case 1 for two reasons. First, in Case 2, Engineer Y took the initiative and contacted several manufacturers to encourage them to produce processes and equipment that will accomplish the same results as those of Engineer X. Second, as an inducement, Engineer Y made a commitment to reference their processes and equipment in future specifications. We believe it is important to evaluate these differences separately to determine whether Engineer Y acted ethically under the facts presented.

We believe that the fact that Engineer Y took the initiative and contacted several manufacturer to encourage them to produce processes and equipment that will accomplish results similar to those of Engineer X's processes and equipment is consistent with the philosophy embodied in the *Code of Ethics*. As we noted in the discussion of Case 1, by taking this initiative Engineer Y is merely using the product developed by Engineer X as a "standard" by which alternative processes and equipment would be measured. There is no indication that Engineer Y is attempting to encourage others to infringe upon a legally obtained patent, but instead Engineer Y appears to be using Engineer X's product as a model or benchmark for the production of a different product that will produce a similar result.

With regard to Engineer Y providing manufacturers with proposed performance specifications, there is no indication that Engineer Y has in any manner infringed upon the patent or other proprietary right of Engineer X. Rather it appears that Engineer Y has developed a set of performance specification based on the performance of Engineer X's processes and equipment and is using those specifications to assist manufacturers in developing similar processes and equipment. So long as Engineer Y is acting consistent with the law, we cannot see how his actions could be condemned as being unethical.

Concerning Engineer Y's inducement by committing to include their processes and equipment in future specifications, we are frankly troubled. It appears that Engineer Y is making an unqualified commitment to specify certain processes and equipment without prior evaluation or review of those processes and equipment. The sole criteria established by Engineer Y to specify the product is that the manufacturer agreed to commit resources to develop the process and equipment alternatives that Engineer Y was seeking. Such a commitment is unethical and unprofessional because Engineer Y is compromising his professional judgment in a manner that could place the interest of his client at risk. By keeping this promise, Engineer Y is in the position of specifying wastewater treatment processes or equipment that is unproven and potentially not of sufficient quality for his client when a higher quality and less costly alternative may have become available on the market. Engineer Y's commitment is clearly the unethical approach in seeking product alternatives to Engineer X's product.

Finally, we note that Engineer Y is ethically obligated to act as faithful agent or trustee of the client. However, as indicated by the facts, Engineer Y has essentially engaged in and independent understanding with a potential vendor to the client. Even if the product is ultimately successful, Engineer Y's actions give the clear appearance of undermining his client's faith and trust, which presumably are part of the basis upon which Engineer Y was selected.

Conclusions

Case 1. It is ethical for Engineer B to use Engineer A's patented processes and equipment as a guide in preparing performance or open specifications.

Case 2. It is not ethical for Engineer Y to endeavor to induce other manufacturers to produce processes and equipment that will perform equally to the patented processes and equipment by making a commitment to include their processes and equipment in his future specifications.

Crediting Work of Another Engineer (BER Case 92-1)

Facts

Engineer A is retained by a city to design a bridge as part of an elevated highway system. Engineer A then retains the services of Engineer B, a structural engineer with expertise in horizontal geometry, superstructure design, and elevations to perform certain aspects of the design services. Engineer B designs the bridge's three curved welded plate girder spans which were critical elements of the bridge design.

Several months following completion of the bridge, Engineer A enters the bridge design into a national organization's bridge design competition. The bridge design wins a prize. However, the entry fails to credit Engineer B for his part of the design.

Question

Was it ethical for Engineer A to fail to give credit to Engineer B for his part in the design?

Discussion

Basic to engineering ethics is the responsibility to issue statements in an objective and truthful manner (Section I.3.). The concept of providing credit for engineering work to those to whom credit is due is fundamental to that responsibility. This is particularly the case where an engineer retains the services of other individuals because the engineer may not possess the education, experience, and expertise to perform the required services for a client. The engineer has an obligation to the client to make this information known (Section II.3.a.). As noted in BER Case 71-l, the principle is not only fair and in the best interests of the profession, but it also recognizes that the professional engineer must assume personal responsibility for his decisions and actions.

In BER Case 71-l, a city department of public works retained Firm A to prepare plans and specifications for a water extension project. Engineer B, chief engineer of the department having authority in such matters, instructed Firm A to submit its plans and specifications without showing the name of the firm on the cover sheets but permitted the firm to show the name of the firm on the working drawings. It was also the policy of the department not to show the name of the design firm in the advertisements for construction bids, in fact, the advertisements stated "plans and specifications as prepared by the city department of public works." The Board noted that the policy of the department is, at best, rather unusual in normal engineering practices and relationships between retained design firms and clients. The Board surmised on the basis of the submitted facts that the department policy was intended to reflect the idea that the plans and specifications when put out to construction bid are those of the department. In concluding that Engineer B acted unethically in adopting and implementing a policy which prohibited the identification of the design firm on the cover sheets for plans and specification, the Board noted that Engineer B, in carrying out the department policy, denied credit to Firm A for its work. The *Code of Ethics* Section III.8.a. states that engineers shall, whenever possible, name the person or persons who may be individually responsible for designs, inventions, writings, or other accomplishments. The Board concluded that under the circumstances, it was possible for Engineer B to name the persons responsible for the design.

While each individual case must be understood based upon the particular facts involved, we believe that Engineer A had an ethical obligation to his client, to Engineer B as well as to the public to take reasonable steps to identify all parties responsible for the design of the bridge.

Conclusion

It was unethical for Engineer A to fail to give credit to Engineer B for his part in the design.

Technical Society Participation (BER Case 82-7)

Facts

Engineer A has been employed by an organization for more than 20 years. During his early years of employment he was encouraged by his superiors to join and participate in the activities of both a technical society and a professional society. Within those societies, Engineer A held several board and committee positions, of which, entry into the key positions was approved by his superiors. He presently holds a committee position.

Engineer A's immediate superior, Engineer B, opposes Engineer A's participation in activities of his professional society on any other than annual leave basis, although existing organization rules encourage the use of excused leave for such purposes. It is Engineer B's view that such participation does not result in "benefits for the employer"; he feels that such participation does not constitute "employee training." Engineer B has refused to permit written communications from Engineer A asking for administrative leave to attend professional society meetings to go through Engineer B to higher level personnel.

When summoned by the chief executive officer (CEO) on another matter, Engineer A took the opportunity to ask his opinion of attendance and participation in technical and professional society meetings by his engineers. The CEO reaffirmed the organization policy.

When Engineer A prepared a travel request to go through his superior, Engineer B, to the CEO, Engineer B refused to forward the travel request and told Engineer A that he did not appreciate Engineer A's going over his head to discuss attendance and participation in technical and professional societies with his superior.

Questions

1. Was it ethical for Engineer A to discuss attendance and participation in technical and professional societies with the CEO without first notifying his superior?
2. Was it ethical for Engineer B to hinder Engineer A's efforts to obtain excused leave in order to attend technical and professional society meetings?

Discussion

It is possible for this Board to review the actions of Engineer A and to conclude that as a factual matter he was disloyal and promoting his own self-interests by going beyond his immediate superior to obtain permission to attend and participate in professional and technical society activities. However, if we were to do so, we would be ignoring the basic underlying philosophy of engineering—professionalism. The essence of professionalism is the unique service a practitioner renders to a client by virtue of having developed special capabilities. In line with that view we believe an employer of engineers has an obligation to treat engineers as professional individuals. It is incumbent on the employer of any employed professional engineer to create an environment conducive to the continued development of professional capabilities. Of course it is the professional obligation of the practitioner to expend some time and effort to continuous expansion of his or her knowledge and capabilities. Such expansion of knowledge may be gained in a variety of ways. We believe one of those ways is by participating in the activities of a professional society. In particular, participation in the committee work of a professional society allows the practicing engineer the opportunity to gain a greater understanding of the new trends and advances in his profession, permits him to interact and exchange views and insights with other engineers, and provides the engineer with a better perspective as to the role of the engineer in society.

We are of the view that a fundamental issue was at stake when Engineer A discussed attendance and participation in technical and professional societies with the CEO. What was at stake was Engineer A's professional integrity and his obligation to expand his knowledge and capabilities.

In addition, we note that it was the general policy of the employer to encourage Engineer A's participation in the activities of technical and professional societies. It was only Engineer A's immediate supervisor, Engineer B, who hindered his efforts to participate. In view of those factors, we are of the view that Engineer A acted professionally and faithfully in his dealings with his employer.

Although it may have been more appropriate for Engineer A to first meet with his supervisor, Engineer B, to inform him of his intention to seek the CEO's permission to attend and participate in the technical and professional organizations' activities, we are not convinced that his failure to do so tended to promote his own self-interest at the expense of the dignity and integrity of the profession. Although his action might be characterized as a deception, given the intransigence of his supervisor, Engineer B, in not permitting him to communicate with his superior on the matter of participation in professional and technical society activities, one can better understand his decision to pursue this route. We find that Engineer A's failure to inform Engineer B of his intention to seek the CEO's permission to attend and participate in technical and professional society activities did not promote his own interest at the expense of the profession.

As for Engineer B, we are of the opinion that his opposition is neither in accord with the *Code* nor supported by experience. Engineers should encourage their employees to participate in a variety of activities in order to foster their professional growth and development. Among these activities are professional and technical society meetings. Engineer B was of the view that Engineer A's participation in technical and professional societies did not constitute "employee training" and did not result in "benefits to the employer." Aside from the question of whether this was in fact an accurate assessment of Engineer A's society activities, there is the issue of whether standards such as "employee training" or "benefits to the employer" are the only yardsticks by which professional and technical society activities and continuing engineering education programs should be measured. We think not but leave that question for another day. It suffices to say that in the instant case, contrary to Engineer B's view, Engineer A's participation in professional and technical society meetings was professionally desirable.

We note, however, that our decision today must not be construed to mean that an engineer should as a matter of course be granted excused leave from his employment without due regard to the needs and requirements of his employer. We believe that Section I.4. mandates that an engineer must be sensitive to the needs and requirements of his employer. When an employer chooses to limit his employees' participation in technical and professional society activity because those employees' services are critical to the operation of his organization, Section I.4. requires the employee to accede to his employer's decision. Although an engineer has an obligation to further his professional growth and development, it should never be pursued in a manner that would be adverse to the interest of his employer.

Conclusions

Q1. It was ethical for Engineer A to discuss attendance and participation in technical and professional societies with the CEO without first notifying his superior.

Q2. It was unethical for Engineer B to hinder Engineer A's efforts to obtain excused leave in order to attend technical and professional society meetings.

INTERNATIONAL ETHICS

Gifts to Foreign Officials (BER Case 76-6)

Facts

Richard Roe, PE, is president and chief executive officer of an engineering firm which has done overseas assignments in various parts of the world. The firm is negotiating for a contract in a foreign country in which it has not worked previously. Roe is advised by a high-ranking government official of that country that it is established practice for those awarded contracts to make personal gifts to the governmental officials who are authorized to award the contracts, and that such practice is legal in that country. Roe is further advised that while the condition is not to be included in the contract, his failure to make the gifts will result in no further work being awarded to the firm and to expect poor cooperation in performing the first contract. He is further told that other firms have adhered to the local practice in regard to such gifts.

Question

Would it be ethical for Roe to accept the contract and make the gifts as described?

Discussion

On its face, the *Code* is clear and direct to the point. There is no question under the factual situation that the gifts are a direct consideration for securing the work, and there is here no pretense about the intention of the foreign government officials. In Case 60-9 we acted upon a domestic case under Rule 4 of the then-prevailing Rules of Professional Conduct, which was the same as the present Section II.5.b. In that case we dealt with three levels of gifts, involving taking employees of a public agency to luncheon or dinner a few times during the year at an average cost of $5 per person; engineers of an industrial firm receiving cash gifts of from $25 to $100 from salesmen for certain products which may be specified by the engineers; and a consulting engineer giving the chief engineer of a client an automobile of the value of approximately $4000 at the completion of a project as an expression of appreciation for the cooperation of the recipient of the gift. (Realistically, these figures would now be much higher than those amounts on account of inflation since 1960.)

It is now worth repeating the language used at that time to indicate the basic principles which should govern the question:

> The question of when a gift is intended to or becomes an inducement to influence one's impartial decision, as distinguished from an expression of friendship or a social custom, has remained a perplexing one over the years. No blanket rule covering all situations has been discovered. The size of the gift is usually a material factor, but must be related to the circumstances of the gift. It would hardly be felt a token gift, such as a cigar, a desk calendar, etc., would be prohibited. It has been customary in the business world for friends and business associates to tender such tokens of recognition or appreciation, and 'picking up the tab' at a business luncheon or dinner is commonplace and well accepted in the mores of our society.
>
> Recognizing the difficulties inherent in passing judgment on each instance, we believe the canons and rules state, in substance, that an engineer may neither offer nor receive a gift which is intended to or will influence his independent professional judgment. The full ap-

> plication of this principle requires the impossible—that we read the state of mind of the donor or donee. Therefore, we must apply a criterion which reasonable men might reasonably infer from the circumstances; that the giving or acceptance of the benefit be a matter of 'good taste,' and such that among reasonable men it might not be of a nature which raises suspicions of favoritism.
>
> Applying these general principles to the situations at hand, we think that an occasional free luncheon or dinner, or a Christmas or birthday present when there is a personal relationship is acceptable practice. On the other hand, cash payments to those in a position to influence decisions favorable or unfavorable to the giver are not in good taste and do immediately raise a suspicion that there is an ulterior motive. Likewise, a very expensive gift has a connotation of placing the recipient in a position of obligation.

From those principles we then concluded that the practice in Situation 1 was ethically permissible, but those in Situations 2 and 3 were unethical.

In the case before us we are not told the amount of the proposed gifts, but we take it in context that they would be substantial. Accordingly, they would clearly be a violation of the *Code* if offered in the United States.

The basic issue remaining is whether the flat prohibition in Section II.5.b. applies to work in a foreign country where the laws and customs permit the gifts to government officials. It is worth noting in considering this point that in a different but related context, the NSPE Board of Directors in July 1966 adopted a so-called "When in Rome" clause to permit the submission of tenders for work in foreign countries when such is required by laws, regulations, or practices of the foreign country.

However, after further discussion and debate, the Board of Directors in January 1968 rescinded the "When in Rome" clause from the *Code* as recommended by the Professional Engineers in Private Practice Section, which recognized a division of opinion on the policy but concluded, "... the profession should maintain a 'pure' position on competitive bidding; otherwise our opposition to competitive bidding will be chipped away, piece by piece."

The issue before us is a very current one. In recent months the press has been filled with reports of investigations of charges that certain industrial concerns have made improper gifts of large sums to foreign officials to secure contracts for their products. The defense to this activity has generally been that the companies making such gifts had "no choice," meaning that without such action they would not have been able to secure the contracts because competitors in other countries would have complied with the practice. There may be some appeal to this line of argument from a purely pragmatic standpoint, but it must of necessity fail in the final analysis.

Even though the practice may be legal and accepted in the foreign country, and even though some might argue on pragmatic grounds that United States commercial companies should "go along" to protect the jobs of employees in this country, we cannot accept it for professional services. No amount of rationalization or explanation will change the public reaction that the profession's claim of placing service before profit has been compromised by a practice which is repugnant to the basic principles of ethical behavior under the laws and customs of this country. Even if the "go along" philosophy is accepted as an exception only for foreign work, the result must be a "chipping away" of ethical standards, leading to contention that such conduct should also be accepted in the United States when and if it is argued that such is the local or area practice.

This approach is not dissimilar to the arguments advanced by those who have so recently been revealed as offering financial payments to public officials to influence the award of contracts for architect-engineer services. The rationale was that "We had no choice. Others were doing it, and if we did not we would not be considered." The short answer is that there is a choice—the choice of declining to be drawn into a seamy procedure for self-gain.

We believe that the *Code* must be read on this most basic point of honor and integrity not only literally, but in the spirit of its purpose—to uphold the highest standards of the profession. Anything less is a rationalization which cannot stand the test of placing service to the public ahead of all other considerations.

Conclusion

It would be unethical for Roe to accept the contract and make the gifts as described.

Marketing Agreements (BER Case 78-7)

Facts

John Doe, PE, has been engaged extensively in recent years in a variety of engineering activities in the international market. He determines that on the basis of his experience, familiarity with the special requirements of engineering work in other countries, and personal contacts with officials of certain foreign countries he could better serve the interests of the engineering profession, as well as his own economic interests, by representing United States firms which wish to engage in international engineering and lack a background in the special fields of knowledge required for that purpose, or which do not have the resources to develop the necessary skills to successfully enter that field.

Recognizing the inability of many U.S. firms to commit themselves to a substantial capital outlay to develop their potential in the international market pending the award of a contract, Doe drafts a plan, called a "Marketing Agreement," under which he offers his services to represent U.S. firms interested in obtaining international work. The agreement calls for Doe to provide information and develop contacts within stated geographical areas, to evaluate potential projects for the firms he represents, to coordinate project development, arrange contract terms between the client and the represented firm, and provide such other special services as the represented firms may authorize.

For these services Doe is to be paid a basic fee, the amount of which is to be negotiated on an individual firm basis, a monthly retainer fee of a negotiated amount on an individual firm basis, and a "marketing fee" of a negotiated percentage of the fees actually collected by the firm he represents for projects which were "marketed" by Doe.

Question

Is it ethical for an engineering firm to enter into such a "Marketing Agreement" with Doe?

Reference

NSPE *Code of Ethics*, Section 11(b) prior to July 1976—"He will not pay, or offer to pay, either directly or indirectly, any commission, political contribution, or a gift, or other consideration in order to secure work, exclusive of securing salaried positions through employment agencies."

Discussion

We presume that the present case does not violate federal laws or laws of the country involved. We dealt with a similar situation in Case 77-1, but with certain differences in the facts, and concluded it was not ethical for an engineer to pay a commission to a commercial marketing firm to secure work for him.

In the case before us, it is clear that the "commercial marketing firm" involved is an individual professional engineer offering his services on a commission basis, in part.

In reaching our conclusion in Case 77-1, we noted an earlier decision in Case 62-4 in which a paramount factor was that a sales representative of an engineering firm to be paid on a combined salary-commission basis was not an engineer. In that case it was concluded that the firm could utilize the sales promotion of a nonengineer, provided he did not discuss engineering aspects of the project, and only commented on the commission payment issue that, "...this method of compensation is undesirable since it could lead to loss of confidence by the public in the professional nature of engineering services."

Extending that comment in the 1977 case, however, we said that the use of a commercial marketing firm would offend the *Code of Ethics* because, "...the engineering firm has control over the conduct of an employee, whereas it has little or no control over the conduct of an outside marketing firm which operates on a commercial basis. The danger is thus much enhanced that a commercial marketing firm may more readily in its zeal to earn its compensation engage in conduct which may adversely reflect upon the dignity or honor of the profession."

Three members of the Board of Ethical Review, while signing the opinion in Case 77-1, expressed "additional views" to the effect that, "...in the context of modern business practices as required by our complex society and the increasing number of U.S. firms exporting their technical expertise on a global basis, there is a serious question as to whether the present language of the *Code* is unduly restrictive while offering at best a limited measure of protection of the public interest." The members of the BER subscribing to the additional views suggested that the issue should be reviewed for a possible change in the pertinent *Code* language or concept. To date, however, Section 11(b) of the *Code* has not been revised.

When the prohibition of commission fees as a basis to secure work is read in conjunction with other parts of Section 11(b), i.e., political contributions or gifts, it would appear that the original purpose and intent were to foreclose circumstances which might arouse doubt or suspicion of impropriety in securing engineering assignments.

It is clear, however, that Section 11(b) prohibits the payment of "any" commission in order to secure work (other than salaried positions), thereby ruling out the permissibility of a commission basis coupled with definite sums as a retainer fee or basic fee.

Conclusion

It is not ethical for an engineering firm to enter into such a "Marketing Agreement" with Doe.

BER Note

All members of the board concur in the "Additional Views" expressed in Case 77-1 and urge revision of the *Code* along those lines. For the information of the reader those additional views were:

"The...members of BER take this opportunity to express additional views on the issues described in Case 77-1, the conclusions of which were unanimously agreed to by board members present.

"It should be understood by all that the BER is constrained to decide cases on the basis of the wording of the *Code of Ethics* then in effect. In some instances the language of the *Code* is broad enough to require an interpretation as to its applicability in a specific situation; however, there is no doubt that a literal reading of Section 11(b) makes the employment of a commercial marketing company and the payment of a commission to secure work a 'per se' violation of this section of the *Code*. Hence the unanimous opinion of BER members on Case 77-1 as described.

"It is axiomatic that the principal thrust of the NSPE *Code of Ethics* is the advancement and protection of the public welfare. Viewed in the context of modern business practices as required by our complex society and the increasing number of U.S. firms exporting their technical expertise on a global basis, there is a serious question as to whether the present language of the *Code* is unduly restrictive while offering at best a limited measure of protection of the public interest. There are many circumstances where it becomes desirable, if not actually necessary, for engineering firms to employ outside services to assist in obtaining work. Marketing firms specializing in the representation of professional engineering firms may well be controlled by qualified professionals fully aware of the proscriptions involved in the marketing of professional services, thus it appears unreasonable to assume that these firms would be unduly susceptible to improper conduct in their zeal to earn their compensation. In any event, clearly it should be the responsibility of the engineering firm to assure that the marketing group which has been authorized to represent its interests does so in a proper manner.

"The payment of a 'commission' based on actual sales vis-à-vis a specified payment schedule or salary if a bona fide employee of the consulting firm presents a more complex issue. It is apparent that the pressures and temptations inherent in this type of financial arrangement could '...lead to a loss of confidence by the public in the professional nature of engineering services,' as stated in a previous case. There are many engineers, however, who feel that the financial arrangement between the engineer and his employees or representatives is a business consideration not directly relevant to ethical considerations.

"It would be well for the NSPE Ethical Practices Committee to review the issues raised by Case 77-1 in the light of current conditions of practice."

Author's Note: Subsequent to the ruling in Case 78-7, the NSPE *Code of Ethics* was revised. Section 11(b) was replaced with a new Section II.5.b which authorizes payments "...to a bona fide employee or bona fide established commercial or marketing agencies retained by them." Presumably, under the revised *Code*, the ruling of the Board of Ethical Review would be different.

RESEARCH ETHICS

Joint Authorship of Technical Paper (BER Case 85-1)

Facts

Engineer A and Engineer B are faculty members at a major university. As part of the requirement for obtaining tenure at the university, both Engineer A and Engineer B are required to author articles for publication in scholarly and technical journals. During Engineer A's years as a graduate student he had developed a paper which was never published and which forms the basis of what he thinks would be an excellent article for publication in a journal. Engineer A discusses his idea with Engineer B and they agree to collaborate in developing the article. Engineer A, the principal author, rewrites the article, bringing it up to date. Engineer B's contributions are minimal. Engineer A agrees to include Engineer B's name as coauthor of the article as a favor in order to enhance Engineer B's chances of obtaining tenure. The article is ultimately accepted and published in a refereed journal.

Questions

1. Was it ethical for Engineer A to use a paper he developed at an earlier time as the basis for an updated article?
2. Was it ethical for Engineer B to accept credit for development of the article?
3. Was it ethical for Engineer A to include Engineer B as co-author of the article?

Discussion

This case presents three distinct issues which, although not directly addressed by the *Code of Ethics* nor earlier BER decisions, are extremely important in regard to the integrity and honesty of intellectual work performed by university engineering faculty.

The first issue relates to that of engineering faculty using material from previous work performed and modifying that material in order to satisfy a requirement to publish. This development has occurred in recent years as a result of the emphasis placed by various universities and colleges upon the importance of publication. With pressures being exerted upon faculty to write articles acceptable for publication, some faculty, as a result of time pressures and other factors, have sometimes "cut corners" in order to satisfy the requirement to publish.

While we stress the importance of performing new and innovative engineering research, we are not convinced that previous work of a high quality could not form the basis of updated research by engineering faculty. Quite often engineering students and faculty embark upon areas of research, and owing to a variety of factors, many beyond their control (time constraints, priorities, funding, etc.), make the decision to postpone the research being conducted. Later, for a number of reasons, they may decide to resume the research. Flowing out of the concluded research may be articles or reports suitable for publication in technical journals. As long as an article is properly updated and the data verified and scrutinized in view of the time lapse, we are of the view that such publication would be entirely proper and ethical.

It may be suggested that because the earlier research was performed not as a faculty member but as an engineering student, the research was performed outside of the scope of the faculty member's current employment and therefore should not be credited as research performed as faculty for the purpose of tenure. We have trouble accepting such an inflexible view, particularly in view of the aforementioned variables that may impact upon the ability to perform research. We think the better course to take is to examine the relative quality of the individual's research rather than to question the chronology of the research. As long as the research is of a high-quality nature, we are satisfied that no ethical violation exists. In view of the fact that the article was brought up to date and was ultimately published in a refereed journal, we are convinced that no ethical problem has emerged.

Turning to a second issue in this case, as noted earlier, we are sensitive to the extremely difficult position in which many faculty members have been placed with regard to the so-called rule of "publish or perish." This Board finds it extremely difficult to sanction a situation whereby Engineer A permits Engineer B, for whatever reason, to share joint authorship on an article when it is clear that Engineer B's contributions to the article are minimal. We think that Section III.3.c. speaks to this point. This Board cannot excuse the conduct of a faculty member who "takes the easy way out" and seeks credit for an article that he did not author. The only way a faculty tenure committee can effectively evaluate tenure candidates is to examine the candidates' qualifications and not the qualifications of someone else. For this Board to decide otherwise would be to sanction a practice entirely at odds with academic honesty and professional integrity. (See Section III.1.)

Finally, the facts of the case raise the question of Engineer A's ethical conduct in agreeing to include Engineer B as co-author of the article as a favor in order to enhance Engineer B's chances of obtaining tenure. However genuine Engineer A's motives may have been under the circumstances, we unqualifiedly reject the action of Engineer A. By permitting Engineer B to misrepresent his achievements in this way, Engineer A has compromised his honesty and forfeited his integrity. Engineer A is unquestionably diminished by this action.

While this Board is fervent in its view and wishes to stress the importance of those three points, we also feel compelled to acknowledge that certain "gray areas" do exist. Frequently, technical articles are written that contain the names of many authors or contributors. Often it is difficult to identify in an objective manner the qualitative contributions of the various authors identified. While we recognize that this practice is a proper means of accurately identifying actual authors contributing to an article, we tend to be somewhat skeptical in general of this practice. We recognize the importance of collaboration in academic endeavors; however, we think that the collaborative effort should produce and reflect a high-quality product worthy of joint authorship, and should not merely be a means by which engineering faculty expand their list of achievements.

Conclusions

Q1. It was ethical for Engineer A to use a paper he developed at an earlier time as the basis for an updated article.

Q2. It was unethical for Engineer B to accept credit for development of the article.

Q3. It was unethical for Engineer A to include Engineer B as co-author of the article.

Credit for Research Work (BER Case 92-7)

Facts

The XYZ Company headed by Engineer A offered to provide funding to professors in the chemistry department of a major university for research on removing poisonous heavy metals (copper, lead, nickel, zinc, chromium) from waste streams. The university then agreed to contract with XYZ company to give the company exclusive use of the technology developed in the field of water treatment and wastewater stream treatment. Under the agreement, XYZ Company will provide a royalty to the university from profits derived from the use of the technology. Also, a group of the university professors organized QRS, a separate company to exploit applications of the technology other than the treatment of water and wastewater.

At the same time that the university research was being conducted, XYZ continued to conduct research in the same area. Performance figures and conclusions were developed. XYZ freely shared the figures and conclusions with QRS organized by the university professors.

At the university, Engineer B, a professor of civil engineering wanted to conduct research and develop a paper relating to the use of the technology to treat sewage. Engineer B contacted the professors in the university's chemistry department. The chemistry professors provided XYZ's data to Engineer B for use in the research and paper. The professors did not reveal to Engineer B that the data was generated by Engineer A and XYZ company.

Engineer B's paper was published in a major journal. Engineer A's data was displayed prominently in the paper and the work of XYZ constituted a major portion of the journal. The paper credits two of the chemistry professors as major authors along with Engineer B. No credit was given to Engineer A or XYZ as the source of the data, the funds that supported the research. After publication Engineer B learns about the actual source of the data and its finding.

Question

Does Engineer B have an obligation under the *Code of Ethics* to clarify the source of the data contained in the paper?

Discussion

The issue of providing credit for research work performed by others is a vital matter in this day and age. Its importance is more than merely crediting contributions of individuals who have performed work in an area of engineering and scientific research. In actual fact, funding decisions for research and development of various technologies are vitally affected by the credit and acknowledgments.

Over the years, the Board has examined these issues in a variety of contexts. In BER Case 75-ll, Engineer A performed certain research and then prepared a paper on an engineering subject based on that research which was duly published in an engineering magazine under his byline. Subsequently, an article on the same subject under the name of Engineer B appeared in another engineering magazine. A substantial portion of the text of Engineer B's article was identified word-for-word with the article authored by Engineer A. Engineer A contacted Engineer B and requested an explanation. Engineer B replied that he had submitted with his article a list of six references, one of which identified the article by Engineer A, but that the list of references had been inadvertently omitted by the editor. He offered his apology to Engineer A for the mishap because his reference credit was not published as intended. In ruling that Engineer B did not act ethically by his actions, we distinguished research from plagiarism. We offered that a "quotation from many sources is research" and "quotation from a single or limited number of sources is plagiarism." However, in either event, it is contemplated that the author will identify and give credit to his sources, single or many. In addition, we noted that the important belief of Engineer B that he would have been without fault if the list of references had been published at the end of the article. This belief represented a lack of understanding of the requirements of the *Code*. Merely listing the work of Engineer A in a list of references to various articles only tells the reader that Engineer B had consulted and read those cited articles of other authors. It no way tells the reader that a large portion of his text is copied from the work of another.

More recently, in BER Case 83-3, Engineer B submitted a proposal to a county council following an interview concerning a project. The proposal included technical information and data that the council requested as a basis for the selection. Smith, a staff member of the council, made Engineer B's proposal available to Engineer A. Engineer A used Engineer B's proposal without Engineer B's consent in developing another proposal, which was subsequently submitted to the council. The extent to which Engineer A used Engineer B's information and data is in dispute between the parties. In finding that it was unethical for Engineer A to use Engineer B's proposal without Engineer B's consent, we indicated that Engineer A had an obligation to refuse to accept the proposal from Smith and also noted that Engineer A's actions constituted unfair competition by improper and questionable methods in violation of the *Code*.

Taking BER Cases 75-ll and 83-3 together, we believe that the instant case can be distinguished from the two earlier cases. Unlike the facts in BER Cases 75-1 l and 83-3, Engineer B did not knowingly fail to credit Engineer A or XYZ corporation for its contributions to the research which formed the basis of his paper. Instead, Engineer B assumed that the material he received from the other professors was developed solely by those professors.

We conclude that Engineer B did not knowingly and deliberately fail to credit Engineer A or XYZ for its contributions to the research. However, we believe that had Engineer B made more of an effort to substantiate the sources contained in his paper, he may have been able to identify those sources. We would also emphasize our deep concern over the conduct of the chemistry professors who for whatever reason(s) mislead Engineer B by failing to reveal the sources of the data. While not technically covered by this *Code*, the conduct of the chemistry professors is clearly deplorable and is unacceptable under the philosophical standards embodied in the *Code of Ethics*.

Finally, we would suggest that Engineer B prepare and request that the journal publish a clarification of the matter explaining how the matter occurred along with an apology for any misunderstanding which may have arisen as a result of the publication of the paper.

Conclusion

Engineer B has an obligation to request that the journal publish a clarification of the matter explaining how the matter occurred along with an apology for any misunderstanding which may have arisen as a result of the publication of the paper.

ADDITIONAL BER CASES

This chapter, of necessity, presents a limited number of cases from the NSPE Board of Ethical Review. As is stated at the beginning of this chapter, the opinions expressed in these cares are based upon data submitted to the BER and do not necessarily represent all of the pertinent facts when applied to any specific case. These opinions are for educational purposes only and should not be construed as expressing any opinion on the ethics of specific individuals.

The seven volume *Opinions of the Board of Ethical Review* contains hundreds of additional case studies. It is available from NSPE Product Fulfillment, PO Box 1020, Sewickley, PA 15143-1020, FAX: (412) 741-0609, Phone Toll Free: (800) 417-0348, E-mail: nspeorders@abdintl.com.

The BER is educational rather than disciplinary, and the cases reviewed and reported in these volumes do not name the actual individuals and firms. These publications are offered to the profession to teach engineering ethics to students and provide instruction to practicing engineers.

APPENDIX

Engineering Codes of Ethics

This Appendix includes six codes of ethics and ethics policies which have been adopted by engineering and engineering-related professional associations. The six documents differ in wording and specific details, yet they all reflect the same ethical principles. The six codes are:

1. American Association of Engineering Societies *Model Guide for Professional Conduct.* AAES is an umbrella organization that includes many different engineering societies among its members.
2. National Society of Professional Engineers *Code of Ethics for Engineers*. NSPE is a professional society that is concerned with issues that affect all engineers regardless of discipline. It is an individual membership organization with affiliated societies in each U.S. state and territory and local chapters in most major cities.
3. Project Management Institute *Code of Ethics for the Project Management Profession.* PMI is a multi-disciplinary organization composed of project management practitioners in a multitude of professions including engineering and construction. It is an individual membership organization.
4. AACE International, the Association for the Advancement of Cost Engineering, *Canon of Ethics.* AACE is a specialty society that is concerned with problems of cost estimation, cost control, planning and scheduling, and project management in engineering. It is an individual membership organization.
5. American Society of Civil Engineers *Code of Ethics.* ASCE is an individual membership disciplinary organization and is one of the original U.S. engineering societies. Its membership totals approximately 125,000 engineers engaged in the practice of civil engineering.
6. Institute of Electrical and Electronics Engineers *Code of Ethics.* IEEE is the largest individual membership engineering society in the U.S. Its membership is approximately 320,000 engineers engaged in the practice of electrical engineering and related fields.

MODEL GUIDE FOR PROFESSIONAL CONDUCT OF THE AMERICAN ASSOCIATION OF ENGINEERING SOCIETIES

Preamble

Engineers recognize that the practice of engineering has a direct and vital influence on the quality of life for all people. Therefore engineers should exhibit high standards of competency, honesty and impartiality; be fair and equitable; and accept a personal responsibility for adherence to applicable laws, the protection of the public health, and maintenance of safety in their professional actions and behavior. These principles govern professional conduct in serving the interests of the public, clients, employers, colleagues and the profession.

The Fundamental Principle

The engineer as a professional is dedicated to improving competence, service, fairness and the exercise of well-founded judgment in the practice of engineering for the public, employers and clients with fundamental concern for the public health and safety in the pursuit of this practice.

Canons of Professional Conduct

Engineers offer services in the areas of their competence and experience, affording full disclosure of their qualifications.

Engineers consider the consequences of their work and societal issues pertinent to it and seek to extend public understanding of those relationships.

Engineers are honest, truthful and fair in presenting information and in making public statements reflecting on professional matters and their professional role.

Engineers engage in professional relationships without bias because of race religion, sex, age, national origin or handicap.

Engineers act in professional matters for each employer or client as faithful agents or trustees, disclosing nothing of a proprietary nature concerning the business affairs or technical processes of any present or former client or employer without specific consent.

Engineers disclose to affected parties known or potential conflicts of interest or other circumstances which might influence—or appear to influence—judgment or impair the fairness or quality of their performance.

Engineers are responsible for enhancing their professional competence throughout their careers and for encouraging similar actions by their colleagues.

Engineers accept responsibility for their actions; seek and acknowledge criticism of their work; offer honest criticism of the work of others; properly credit the contributions of others; and do not accept credit for work not theirs.

Engineers perceiving a consequence of their professional duties to adversely affect the present or future public health and safety shall formally advise their employers or clients and, if warranted consider further disclosure.

Engineers act in accordance with all applicable laws and the ____________________________[1] rules of conduct, and lend support to others who strive to do likewise.

[1] AAES Member Societies are urged to make reference here to the appropriate code of conduct to which their members will be bound.

Approved by AAES Board of Governors 12/13/84.

NSPE Code of Ethics for Engineers

Preamble

Engineering is an important and learned profession. As members of this profession, engineers are expected to exhibit the highest standards of honesty and integrity. Engineering has a direct and vital impact on the quality of life for all people. Accordingly, the services provided by engineers require honesty, impartiality, fairness and equity, and must be dedicated to the protection of the public health, safety, and welfare. Engineers must perform under a standard of professional behavior that requires adherence to the highest principles of ethical conduct.

I. Fundamental Canons

Engineers, in the fulfillment of their professional duties, shall:

1. Hold paramount the safety, health and welfare of the public.

2. Perform services only in areas of their competence.

3. Issue public statements only in an objective and truthful manner.

4. Act for each employer or client as faithful agents or trustees.

5. Avoid deceptive acts.

6. Conduct themselves honorably, responsibly, ethically, and lawfully so as to enhance the honor, reputation, and usefulness of the profession.

II. Rules of Practice

1. Engineers shall hold paramount the safety, health, and welfare of the public.

a. If engineers' judgment is overruled under circumstances that endanger life or property, they shall notify their employer or client and such other authority as may be appropriate.

b. Engineers shall approve only those engineering documents that are in conformity with applicable standards.

c. Engineers shall not reveal facts, data or information without the prior consent of the client or employer except as authorized or required by law or this *Code*.

d. Engineers shall not permit the use of their name or associate in business ventures with any person or firm that they believe are engaged in fraudulent or dishonest enterprise.

e. Engineers having knowledge of any alleged violation of this *Code* shall report thereon to appropriate professional bodies and, when relevant, also to public authorities, and cooperate with the proper authorities in furnishing such information or assistance as may be required.

2. Engineers shall perform services only in the areas of their competence.

a. Engineers shall undertake assignments only when qualified by education or experience in the specific technical fields involved.

b. Engineers shall not affix their signatures to any plans or documents dealing with subject matter in which they lack competence, nor to any plan or document not prepared under their direction and control.

c. Engineers may accept assignments and assume responsibility for coordination of an entire project and sign and seal the engineering documents for the entire project, provided that each technical segment is signed and sealed only by the qualified engineers who prepared the segment.

3. Engineers shall issue public statements only in an objective and truthful manner.

a. Engineers shall be objective and truthful in professional reports, statements, or testimony. They shall include all relevant and pertinent information in such reports, statements, or testimony, which should bear the date indicating when it was current.

b. Engineers may express publicly technical opinions that are founded upon knowledge of the facts and competence in the subject matter.

c. Engineers shall issue no statements, criticisms, or arguments on technical matters that are inspired or paid for by interested parties, unless they have prefaced their comments by explicitly identifying the interested parties on whose behalf they are speaking, and by revealing the existence of any interest the engineers may have in the matters.

4. Engineers shall act for each employer or client as faithful agents or trustees.

a. Engineers shall disclose all known or potential conflicts of interest that could influence or appear to influence their judgment or the quality of their services.

b. Engineers shall not accept compensation, financial or otherwise, from more than one party for services on the same project, or for services pertaining to the same project, unless the circumstances are fully disclosed and agreed to by all interested parties.

c. Engineers shall not solicit or accept financial or other valuable consideration, directly or indirectly, from outside agents in connection with the work for which they are responsible.

d. Engineers in public service as members, advisors, or employees of a governmental or quasi-governmental body or department shall not participate in decisions with respect to services solicited or provided by them or their organizations in private or public engineering practice.

e. Engineers shall not solicit or accept a contract from a governmental body on which a principal or officer of their organization serves as a member.

5. Engineers shall avoid deceptive acts.

a. Engineers shall not falsify their qualifications or permit misrepresentation of their or their associates' qualifications. They shall not misrepresent or exaggerate their responsibility in or for the subject matter of prior assignments. Brochures or other presentations incident to the solicitation of employment shall not misrepresent pertinent facts concerning employers, employees, associates, joint venturers, or past accomplishments.

b. Engineers shall not offer, give, solicit or receive, either directly or indirectly, any contribution to influence the award of a contract by public authority, or which may be reasonably construed by the public as having the effect of intent to influencing the awarding of a contract. They shall not offer any gift or other valuable consideration in order to secure work. They shall not pay a commission, percentage, or brokerage fee in order to secure work, except to a bona fide employee or bona fide established commercial or marketing agencies retained by them.

III. Professional Obligations

1. Engineers shall be guided in all their relations by the highest standards of honesty and integrity.

a. Engineers shall acknowledge their errors and shall not distort or alter the facts.

b. Engineers shall advise their clients or employers when they believe a project will not be successful.

c. Engineers shall not accept outside employment to the detriment of their regular work or interest. Before accepting any outside engineering employment they will notify their employers.

d. Engineers shall not attempt to attract an engineer from another employer by false or misleading pretenses.

e. Engineers shall not actively participate in strikes, picket lines, or other collective coercive action.

f. Engineers shall not promote their own interest at the expense of the dignity and integrity of the profession.

2. Engineers shall at all times strive to serve the public interest.

a. Engineers shall seek opportunities to participate in civic affairs; career guidance for youths; and work for the advancement of the safety, health and well-being of their community.

b. Engineers shall not complete, sign, or seal plans and/or specifications that are not in conformity with applicable engineering standards. If the client or employer insists on such unprofessional conduct, they shall notify the proper authorities and withdraw from further service on the project.

c. Engineers shall endeavor to extend public knowledge and appreciation of engineering and its achievements.

3. Engineers shall avoid all conduct or practice that deceives the public.

a. Engineers shall avoid the use of statements containing a material misrepresentation of fact or omitting a material fact.

b. Consistent with the foregoing, Engineers may advertise for recruitment of personnel.

c. Consistent with the foregoing, Engineers may prepare articles for the lay or technical press, but such articles shall not imply credit to the author for work performed by others.

4. Engineers shall not disclose, without consent, confidential information concerning the business affairs or technical processes of any present or former client or employer, or public body on which they serve.

a. Engineers shall not, without the consent of all interested parties, promote or arrange for new employment or practice in connection with a specific project for which the Engineer has gained particular and specialized knowledge.

b. Engineers shall not, without the consent of all interested parties, participate in or represent an adversary interest in connection with a specific project or proceeding in which the Engineer has gained particular specialized knowledge on behalf of a former client or employer.

5. Engineers shall not be influenced in their professional duties by conflicting interests.

a. Engineers shall not accept financial or other considerations, including free engineering designs, from material or equipment suppliers for specifying their product.

b. Engineers shall not accept commissions or allowances, directly or indirectly, from contractors or other parties dealing with clients or employers of the Engineer in connection with work for which the Engineer is responsible.

6. Engineers shall not attempt to obtain employment or advancement or professional engagements by untruthfully criticizing other engineers, or by other improper or questionable methods.

a. Engineers shall not request, propose, or accept a commission on a contingent basis under circumstances in which their judgment may be compromised.

b. Engineers in salaried positions shall accept part-time engineering work only to the extent consistent with policies of the employer and in accordance with ethical considerations.

c. Engineers shall not, without consent, use equipment, supplies, laboratory, or office facilities of an employer to carry on outside private practice.

7. Engineers shall not attempt to injure, maliciously or falsely, directly or indirectly, the professional reputation, prospects, practice, or employment of other engineers. Engineers who believe others are guilty of unethical or illegal practice shall present such information to the proper authority for action.

a. Engineers in private practice shall not review the work of another engineer for the same client, except with the knowledge of such engineer, or unless the connection of such engineer with the work has been terminated.

b. Engineers in governmental, industrial, or educational employ are entitled to review and evaluate the work of other engineers when so required by their employment duties.

c. Engineers in sales or industrial employ are entitled to make engineering comparisons of represented products with products of other suppliers.

8. Engineers shall accept personal responsibility for their professional activities, provided, however, that Engineers may seek indemnification for services arising out of their practice for other than gross negligence, where the Engineer's interests cannot otherwise be protected.

a. Engineers shall conform with state registration laws in the practice of engineering.

b. Engineers shall not use association with a nonengineer, a corporation, or partnership as a "cloak" for unethical acts.

9. Engineers shall give credit for engineering work to those to whom credit is due, and will recognize the proprietary interests of others.

a. Engineers shall, whenever possible, name the person or persons who may be individually responsible for designs, inventions, writings, or other accomplishments.

b. Engineers using designs supplied by a client recognize that the designs remain the property of the client and may not be duplicated by the Engineer for others without express permission.

c. Engineers, before undertaking work for others in connection with which the Engineer may make improvements, plans, designs, inventions, or other records that may justify copyrights or patents, should enter into a positive agreement regarding ownership.

d. Engineers' designs, data, records, and notes referring exclusively to an employer's work are the employer's property. Employer should indemnify the Engineer for use of the information for any purpose other than the original purpose.

As Revised July 1996

"By order of the United States District Court for the District of Columbia, former Section 11(c) of the NSPE *Code of Ethics* prohibiting competitive bidding, and all policy statements, opinions, rulings or other guidelines interpreting its scope, have been rescinded as unlawfully interfering with the legal right of engineers, protected under the antitrust laws, to provide price information to prospective clients; accordingly, nothing contained in the NSPE *Code of Ethics*, policy statements, opinions, rulings or other guidelines prohibits the submission of price quotations or competitive bids for engineering services at any time or in any amount."

Statement by NSPE Executive Committee

In order to correct misunderstandings which have been indicated in some instances since the issuance of the Supreme Court decision and the entry of the Final Judgment, it is noted that in its decision of April 25, 1978, the Supreme Court of the United States declared: "The Sherman Act does not require competitive bidding."

It is further noted that as made clear in the Supreme Court decision:

1. Engineers and firms may individually refuse to bid for engineering services.

2. Clients are not required to seek bids for engineering services.

3. Federal, state, and local laws governing procedures to procure engineering services are not affected, and remain in full force and effect.

4. State societies and local chapters are free to actively and aggressively seek legislation for professional selection and negotiation procedures by public agen-

cies.

5. State registration board rules of professional conduct, including rules prohibiting competitive bidding for engineering services, are not affected and remain in full force and effect. State registration boards with authority to adopt rules of professional conduct may adopt rules governing procedures to obtain engineering services.

6. As noted by the Supreme Court, "nothing in the judgment prevents NSPE and its members from attempting to influence governmental action . . ."

NOTE: In regard to the question of application of the *Code* to corporations vis-à-vis real persons, business form or type should not negate nor influence conformance of individuals to the *Code*. The *Code* deals with professional services, which services must be performed by real persons. Real persons in turn establish and implement policies within business structures. The *Code* is clearly written to apply to the Engineer and items incumbent on members of NSPE to endeavor to live up to its provisions. This applies to all pertinent sections of the *Code*.

Code of Ethics for The Project Management Profession

PREAMBLE: Project Management Professionals, in the pursuit of the profession, affect the quality of life for all people in our society. Therefore, it is vital that Project Management Professionals conduct their work in an ethical manner to earn and maintain the confidence of team members, colleagues, employees, employers, clients, and the public.

ARTICLE I: Project Management Professionals shall maintain high standards of personal and professional conduct, and:

a. Accept responsibility for their actions.

b. Undertake projects and accept responsibility only if qualified by training or experience, or after full disclosure to their employers or clients of pertinent qualifications.

c. Maintain their professional skills at the state of the art and recognize the importance of continued personal development and education.

d. Advance the integrity and prestige of the profession by practicing in a dignified manner.

e. Support this code and encourage colleagues and co-workers to act in accordance with this code.

f. Support the professional society by actively participating and encouraging colleagues and co-workers to participate.

g. Obey the laws of the country in which work is being performed.

ARTICLE II: Project Management Professionals shall, in their work:

a. Provide the necessary project leadership to promote maximum productivity while striving to minimize costs.

b. Apply state of the art project management tools and techniques to ensure quality, cost, and time objectives, as set forth in the project plan, are met.

c. Treat fairly all project team members, colleagues, and co-workers, regardless of race, religion, sex, age or national origin.

d. Protect project team members from physical and mental harm.

e. Provide suitable working conditions and opportunities for project team members.

f. Seek, accept, and offer honest criticism of work, and properly credit the contribution of others.

g. Assist project team members, colleagues, and co-workers in their professional development.

ARTICLE III: Project Management Professionals shall, in their relations with employers and clients:

a. Act as faithful agents or trustees for their employers and clients in professional and business matters.

b. Keep information on the business affairs or technical process of an employer or client in confidence while employed, and later, until such information is properly released.

c. Inform their employers, clients, professional societies or public agencies of which they are members or to which they may make any presentations, of any circumstance that could lead to a conflict of interest.

d. Neither give nor accept, directly or indirectly, any gift, payment, or service of more than nominal value to or from those having business relationships with their employers or clients.

e. Be honest and realistic in reporting project quality, cost, and time.

ARTICLE IV: Project Management Professionals shall, in fulfilling their responsibilities to the community:

a. Protect the safety, health, and welfare of the public and speak out against abuses in these areas affecting the public interest.

b. Seek to extend public knowledge and appreciation of the project management profession and its achievements.

This revision was effective immediately upon approval, April 1, 1989.

AACE International Canon of Ethics

Introduction

The AACE member, to uphold and advance the honor and dignity of Cost Engineering and the Cost Management profession and in keeping with the high standards of ethical conduct will (1) be honest and impartial and will serve employer, clients, and the public with devotion; (2) strive to increase the competence and prestige of their profession; and (3) will apply knowledge and skill to advance human welfare.

I. Relations With the Public

A. Members will hold paramount the safety, health, and welfare of the public, including that of future generations.

B. Members will endeavor to extend public knowledge and appreciation of cost engineering and cost management and its achievements, and will oppose any untrue, unsupported, or exaggerated statements regarding cost engineering and cost management.

C. Members will be dignified and modest, ever upholding the honor and dignity of their profession, and will refrain from self-laudatory advertising.

D. Members will express an opinion on a cost engineering or cost management subject only when it is founded on adequate knowledge and honest conviction.

E. On cost engineering or cost management matters, members will issue no statements, criticisms, or arguments that are inspired or paid for by an interested party or parties, unless they preface their comments by identifying themselves, by disclosing the identities of the party or parties on whose behalf they are speaking, and by revealing the existence of any pecuniary interest they may have in matters under discussion.

F. Members will approve or seal only those documents, reviewed or prepared by them, which are determined to be safe for public health and welfare in conformity with accepted cost engineering, cost management and economic standards.

G. Members whose judgment is overruled under circumstances where the safety, health, and welfare of the public are endangered shall inform their clients or employers of the possible consequences.

H. Members will work through professional societies to encourage and support others who follow these concepts.

I. Members will work only with those who follow these concepts.

J. Members shall be objective and truthful in professional reports, statements, or testimony. They shall include all relevant and pertinent information in such reports, statements, and testimony.

II. Relations With Employers and Clients

A. Members will act in all matters as a faithful agent or trustee for each employer or client.

B. Members will act fairly and justly toward vendors and contractors and will not accept any commissions or allowances from vendors or contractors, directly or indirectly.

C. Members will inform their employer or client of financial interest in any potential vendor or contractor, or in any invention, machine, or apparatus that is involved in a project or work for either employer or client. Members will not allow such interest to affect any decisions regarding cost engineering or cost management services that they may be called upon to perform.

D. When, as a result of their studies, members believe a project(s) will not be successful, or if their cost engineering and cost management or economic judgment is overruled, they shall so advise their employer or client.

E. Members will undertake only those cost engineering and cost management assignments for which they are qualified. Members will engage or advise their employers or clients to engage specialists whenever their employer's or client's interests are served best by such an arrangement. Members will cooperate fully with specialists so engaged.

F. Members shall treat information coming to them in the course of their assignments as confidential and shall not use such information as a means of making personal profit if such action is adverse to the interests of their clients, their employers, or the public.

1. Members will not disclose confidential information concerning the business affairs or technical processes of any present or former employer or client or bidder under evaluation, without consent, unless required by law.

2. Members shall not reveal confidential information or finding of any commission or board of which they are members, unless required by law.

3. Members shall not duplicate for others, without express permission of the client(s), designs, calculations, sketches, etc., supplied to them by clients.

4. Members shall not use confidential information coming to them in the course of their assignments as a means of making personal profit if such action is adverse to the interests of their clients, employers, or the public.

G. Members will not accept compensation—financial or otherwise—from more than one party for the same service, or for other services pertaining to the same work, without the consent of all interested parties.

H. Employed members will engage in supplementary employment or consulting practice only with the consent of their employer.

I. Members shall not use equipment, supplies, laboratory, or office facilities of their employers to carry on outside private practice without the consent of their employers.

J. Members shall not solicit a contract from a governmental body on which a principal officer or employee of their organization serves as a member.

K. The member shall act with fairness and justice to all parties when administering a construction (or other) contract.

L. Before undertaking work for others in which the member may make improvements, plans, designs, inventions, or records that may justify copyrights or patents, the member shall enter into a positive agreement regarding the rights of respective parties.

M. Members shall admit and accept their own errors when proven wrong and refrain from distorting or altering the facts to justify their decisions.

N. Members shall not attempt to attract an employee from another employer by false or misleading representations.

O. Members shall act in professional matters for each employer or client as faithful agents or trustees and shall avoid conflicts of interest.

1. Members shall avoid all known or potential conflicts of interest with their employers or clients and shall promptly inform their employers or clients of any business association, interests, or circumstances that could influence their judgment or the quality of their services.

2. Members shall not solicit or accept gratuities, directly or indirectly, from contractors, their agents, or other parties dealing with their clients or employers in connection with work for which they are responsible.

III. Relations With Other Professionals

A. Members will take care that credit for cost engineering and cost management work is given to those to whom credit is properly due.

B. Members will provide prospective employees with complete information on working conditions and their proposed status of employment. After employment begins, they will keep the employee informed of any changes in status and working conditions.

C. Members will uphold the principle of appropriate and adequate compensation for those engaged in cost engineering and cost management work, including those in subordinate capacities.

D. Members will endeavor to provide opportunity for the professional development and advancement of individuals in their employ or under their supervision.

E. Members will not attempt to supplant other cost engineers or cost management professionals in a particular employment after becoming aware that definite steps have been taken toward the others' employment or after they have been employed.

F. Members shall not maliciously or falsely, directly or indirectly, injure the professional reputation, prospects, practice, or employment of another, nor shall they indiscriminately criticize another's work. Proof that another cost professional has been unethical, illegal, or unfair in his/her practice shall be cause for advising the proper authority.

G. Members will not compete unfairly with other cost professionals.

H. Members will cooperate in advancing the cost engineering and cost management profession by interchanging information and experience with other cost professionals and students, by contributing to public communication media and to cost engineering, cost management and scientific societies and schools.

I. Members will not request, propose, or accept professional commissions on a contingent basis under circumstances that compromise their professional judgments.

J. Members will not falsify or permit misrepresentation of their own or their associates' academic or professional qualifications. They shall not misrepresent or exaggerate their degrees or responsibility in or for the subject matter of prior assignments. Brochures or other presentations incident to the solicitation of employment shall not misrepresent pertinent facts concerning employers, employees, associates, joint ventures, accomplishments, or membership in technical societies.

K. Members will prepare articles for the lay or technical press that are only factual, dignified, and free from ostentatious or laudatory implications. Such articles shall not imply credit to the cost professionals for other than their direct participation in the work described unless credit is given to others for their share of the work.

L. Members will not campaign, solicit support, or otherwise coerce other cost professionals to support their candidacy or the candidacy of a colleague for elective office in a technical association.

IV. Standards of Professional Performance

A. Members shall be dignified and modest in explaining their work and merit and will avoid any act tending to promote their own interests at the expense of the integrity, honor, and dignity of the profession.

B. Members, when serving as expert witnesses, shall express a cost engineering and cost management opinion only when it is founded upon adequate knowledge of the facts, upon a background of technical competence, and upon honest conviction.

C. Members shall continue their professional development throughout their careers and shall provide opportunities for the professional development of those cost professionals under their supervision.

1. Members should keep current in their specialty fields by engaging in professional practice, participating in continuing education courses, reading in the technical literature, and attending professional meetings and seminars.

2. Members should encourage their cost engineering and cost management employees to become certified at the earliest possible date.

3. Members should encourage their cost engineering and cost management employees to attend and present papers at professional and technical society meetings.

4. Members shall uphold the principle of mutually satisfying relationships between employers and employees with respect to terms of employment including professional grade descriptions, salary ranges, and fringe benefits.

CODE OF ETHICS*

Effective January 1, 1977

Fundamental Principles**

Engineers uphold and advance the integrity, honor and dignity of the engineering profession by:

1. *using their knowledge and skill for the enhancement of human welfare and the environment;*

2. *being honest and impartial and serving with fidelity the public, their employers and clients;*

3. *striving to increase the competence and prestige of the engineering profession; and*

4. *supporting the professional and technical societies of their disciplines.*

Fundamental Canons

1. *Engineers shall hold paramount the safety, health and welfare of the public and shall strive to comply with the principles of sustainable development in the performance of their professional duties.*

2. *Engineers shall perform services only in areas of their competence.*

3. *Engineers shall issue public statements only in an objective and truthful manner.*

4. *Engineers shall act in professional matters for each employer or client as faithful agents or trustees, and shall avoid conflicts of interest.*

5. *Engineers shall build their professional reputation on the merit of their services and shall not compete unfairly with others.*

6. *Engineers shall act in such a manner as to uphold and enhance the honor, integrity, and dignity of the engineering profession.*

7. *Engineers shall continue their professional development throughout their careers, and shall provide opportunities for the professional development of those engineers under their supervision.*

Guidelines to Practice Under the Fundamental Canons of Ethics

CANON 1. Engineers shall hold paramount the safety, health and welfare of the public and shall strive to comply with the principles of sustainable development in the performance of their professional duties.

a. Engineers shall recognize that the lives, safety, health and welfare of the general public are dependent upon engineering judgments, decisions and practices incorporated into structures, machines, products, processes and devices.

b. Engineers shall approve or seal only those design documents, reviewed or prepared by them, which are determined to be safe for public health and welfare in conformity with accepted engineering standards.

c. Engineers whose professional judgment is overruled under circumstances where the safety, health and welfare of the public are endangered, or the principles of sustainable development ignored, shall inform their clients or employers of the possible consequences.

d. Engineers who have knowledge or reason to believe that another person or firm may be in violation of any of the provisions of Canon 1 shall present such information to the proper authority in writing and shall cooperate with the proper authority in furnishing such further information or assistance as may be required.

e. Engineers should seek opportunities to be of constructive service in civic affairs and work for the advancement of the safety, health and well-being of their communities, and the protection of the environment through the practice of sustainable development.

f. Engineers should be committed to improving the environment by adherence to the principles of sustainable development so as to enhance the quality of life of the general public.

CANON 2. Engineers shall perform services only in areas of their competence.

a. Engineers shall undertake to perform engineering assignments only when qualified by education or experience in the technical field of engineering involved.

b. Engineers may accept an assignment requiring education or experience outside of their own fields of competence, provide their services are restricted to those phases of the project in which they are qualified. All other phases of such project shall be performed by qualified associates, consultants, or employees.

c. Engineers shall not affix their signatures or seals to any engineering plan or document dealing with subject matter in which they lack competence by virtue of education or experience or to any such plan or document not reviewed or prepared under their supervisory control.

CANON 3. Engineers shall issue public statements only in an objective and truthful manner.

a. Engineers should endeavor to extend the public knowledge of engineering and sustainable development, and shall not participate in the dissemination of untrue, unfair or exaggerated statements regarding engineering.

b. Engineers shall be objective and truthful in professional reports, statements, or testimony. They shall include all relevant and pertinent information in such reports, statements, or testimony.

c. Engineers, when serving as expert witnesses, shall express an engineering opinion only when it is founded upon adequate knowledge of the facts, upon a background of technical competence, and upon honest conviction.

d. Engineers shall issue no statements, criticisms, or arguments on engineering matters which are inspired or paid for by interested parties, unless they indicate on whose behalf the statements are made.

e. Engineers shall be dignified and modest in explaining their work and merit, and will avoid any act tending to promote their own interests at the expense of the integrity, honor and dignity of the profession.

CANON 4. Engineers shall act in professional matters for each employer or client as faithful agents or trustees, and shall avoid conflicts of interest.

a. Engineers shall avoid all known or potential conflicts of interest with their employers or clients and shall promptly inform their employers or clients of any business association, interests, or circumstances which could influence their judgment or the quality of their services.

b. Engineers shall not accept compensation from more than one party for services on the same project, or for services pertaining to the same project, unless the circumstances are fully disclosed to and agreed to, by all interested parties.

c. Engineers shall not solicit or accept gratuities, directly or indirectly, from contractors, their agents, or other parties dealing with their clients or employers in connection with work for which they are responsible.

d. Engineers in public service as members, advisors, or employees of a governmental body or department shall not participate in considerations or actions with respect to services solicited or provided by them or their organization in private or public engineering practice.

e. Engineers shall advise their employers or clients when, as a result of their studies, they believe a project will not be successful.

f. Engineers shall not use confidential information coming to them in the course of their assignments as a means of making personal profit if such action is adverse to the interests of their clients, employers or the public.

g. Engineers shall not accept professional employment outside of their regular work or interest without the knowledge of their employers.

CANON 5. Engineers shall build their professional reputation on the merit of their services and shall not compete unfairly with others.

a. Engineers shall not give, solicit or receive either directly or indirectly, any political contribution, gratuity, or unlawful consideration in order to secure work, exclusive of securing salaried positions through employment agencies.

b. Engineers should negotiate contracts for professional services fairly and on the basis of demonstrated competence and qualifications for the type of professional service required.

c. Engineers may request, propose or accept professional commissions on a contingent basis only under circumstances in which their professional judgments would not be compromised.

d. Engineers shall not falsify or permit misrepresentation of their academic or professional qualifications or experience.

e. Engineers shall give proper credit for engineering work to those to whom credit is due, and shall recognize the proprietary interests of others. Whenever possible, they shall name the person or persons who may be responsible for designs, inventions, writings or other accomplishments.

Engineers may advertise professional services in a way that does not contain misleading language or is in any other manner derogatory to the dignity of the profession. Examples of permissible advertising are as follows:

Professional cards in recognized, dignified publications, and listings in rosters or directories published by responsible organizations, provided that the cards or listings are consistent in size and content and are in a section of the publication regularly devoted to such professional cards.

Brochures which factually describe experience, facilities, personnel and capacity to render service, providing they are not misleading with respect to the engineer's participation in projects described.

Display advertising in recognized dignified business and professional publications, providing it is factual and is not misleading with respect to the engineer's extent of participation in projects described.

A statement of the engineers' names or the name of the firm and statement of the type of service posted on projects for which they render services.

Preparation or authorization of descriptive articles for the lay or technical press, which are factual and dignified. Such articles shall not imply anything more than direct participation in the project described.

Permission by engineers for their names to be used in commercial advertisements, such as may be published by contractors, material suppliers, etc., only by means of a modest, dignified notation acknowledging the engineers' participation in the project described. Such permission shall not include public endorsement of proprietary products.

f. Engineers shall not maliciously or falsely, directly or indirectly, injure the professional reputation, prospects, practice or employment of another engineer or indiscriminately criticize another's work.

g. Engineers shall not use equipment, supplies, laboratory or office facilities of their employers to carry on outside private practice without the consent of their employers.

CANON 6. Engineers shall act in such a manner as to uphold and enhance the honor, integrity, and dignity of the engineering profession.

a. Engineers shall not knowingly act in a manner which will be derogatory to the honor, integrity, or dignity of the engineering profession or knowingly engage in business or professional practices of a fraudulent, dishonest or unethical nature.

CANON 7. Engineers shall continue their professional development throughout their careers, and shall provide opportunities for the professional development of those engineers under their supervision.

a. Engineers should keep current in their specialty fields by engaging in professional practice, participating in continuing education courses, reading in the technical literature, and attending professional meetings and seminars.

b. Engineers should encourage their engineering employees to become registered at the earliest possible date.

c. Engineers should encourage engineering employees to attend and present papers at professional and technical society meetings.

d. Engineers shall uphold the principle of mutually satisfying relationships between employers and employees with respect to terms of employment including professional grade descriptions, salary ranges, and fringe benefits.

*As adopted September 25, 1976 and amended October 25, 1980, April 17, 1993 and November 10, 1996.

**The American Society of Civil Engineers adopted THE FUNDAMENTAL PRINCIPLES of the ABET Code of Ethics of Engineers as accepted by the Accreditation Board for Engineering and Technology, Inc. (ABET)

IEEE Code of Ethics

We, the members of the IEEE, in recognition of the importance of our technologies in affecting the quality of life throughout the world, and in accepting a personal obligation to our profession, its members and the communities we serve, do hereby commit ourselves to the highest ethical and professional conduct and agree:

1 to accept responsibility in making engineering decisions consistent with the safety, health and welfare of the public, and to disclose promptly factors that might endanger the public or the environment;

2 to avoid real or perceived conflicts of interest whenever possible, and to disclose them to affected parties when they do exist;

3 to be honest and realistic in stating claims or estimates based on available data;

4 to reject bribery in all its forms;

5 to improve the understanding of technology, its appropriate application, and potential consequences;

6 to maintain and improve our technical competence and to undertake technological tasks for others only if qualified by training or experience, or after full disclosure of pertinent limitations;

7 to seek, accept, and offer honest criticism of technical work, to acknowledge and correct errors, and to credit properly the contributions of others;

8 to treat fairly all persons regardless of such factors as race, religion, gender, disability, age, or national origin;

9 to avoid injuring others, their property, reputation, or employment by false or malicious action;

10 to assist colleagues and co-workers in their professional development and to support them in following this code of ethics.

Approved by the IEEE Board of Directors

August 1990

Index

About the Author

KENNETH K. HUMPHREYS, P.E., C.C.E. is a Consulting Engineer, Granite Falls, North Carolina. Previously the Executive Director of AACE International (formerly the American Association of Cost Engineers), Morgantown, Virginia, he is recognized as a Certified Cost Engineer by AACE International, the Mexican Society of Cost and Economic Engineers, and the International Cost Engineering Council. Dr. Humphreys is the author or coauthor of over 200 professional papers, and the coauthor or coeditor of *Project and Cost Engineers' Handbook, Third Edition, Revised and Expanded* (with Lloyd M. English), *Effective Project Management Through Applied Cost and Schedule Control (with James A. Bent),* and *Basic Cost Engineering, Third Edition, Revised and Expanded* (with Paul Wellman) [All titles, Marcel Dekker, Inc.]. A registered professional engineer, Dr. Humphreys is a Fellow of AACE International and the Association of Cost Engineers (United Kingdom), and a member of the National Society of Professional Engineers, among others. Dr. Humphreys received the B.S. degree (1959) in chemical engineering from the Carnegie Institute of Technology, Pittsburgh, Pennsylvania, the M.S. degree (1968) in materials science engineering from West Virginia University, Morgantown, and Ph.D. degree (1990) in engineering from Kennedy-Western University, Agoura Hills, California.

LaVergne, TN USA
18 March 2010

176479LV00006B/73/P

9 780824 782085